A REASSESSMENT OF THE MARINE SALVAGE POSTURE OF THE UNITED STATES

COMMITTEE ON MARINE SALVAGE ISSUES
MARINE BOARD
COMMISSION ON ENGINEERING AND TECHNICAL SYSTEMS
NATIONAL RESEARCH COUNCIL

National Academy Press
Washington, D.C. 1994

NOTICE: The project that is the subject of this report was approved by the Governing Board of the National Research Council, whose members are drawn from the councils of the National Academy of Sciences, the National Academy of Engineering, and the Institute of Medicine. The members of the panel responsible for the report were chosen for their special competencies and with regard for appropriate balance.

This report has been reviewed by a group other than the authors according to procedures approved by a Report Review Committee consisting of members of the National Academy of Sciences, the National Academy of Engineering, and the Institute of Medicine.

The program described in this report is supported by cooperative agreement No. 14-35-0001-30475 between the Minerals Management Service of the U.S. Department of the Interior and the National Academy of Sciences and by interagency cooperative agreement No. DTMA91-94-G-00003 between the Maritime Administration of the Department of Transportation and the National Academy of Sciences.

Library of Congress Catalog Card Number 94-68622
International Standard Book Number 0-309-05149-5

Limited copies are available from:

Marine Board
Commission on Engineering and
Technical Systems
National Research Council
2101 Constitution Avenue
Washington, D.C. 20418

Additional copies are for sale from:

National Academy Press
2101 Constitution Avenue
Box 285
Washington, D.C. 20055
800-624-6242 or 202-334-3313 (in the Washington Metropolitan Area)

B-472

Printed in the United States of America

Cover: The *Argo Merchant*, photograph courtesy of the U.S. Coast Guard

Committee on Marine Salvage Issues

MEMBERS

GORDON W. PAULSEN, *Chairman*, Healy & Baillie, New York, New York
PETER F. BONTADELLI, Office of Oil Spill Prevention & Response, California Department of Fish and Game, Sacramento
J. HUNTLY BOYD, JR., Booz-Allen, Hamilton, Inc., Arlington, Virginia
KENNETH J. FULLWOOD, Maritime Relations and Environmental Affairs, Safety and Nautical Services, Mobil Shipping and Transportation Company, Fairfax, Virginia
RICHARD F. LEE, Skidaway Institute of Oceanography, Savannah, Georgia
J.H. (Mick) LEITZ, J.H. Leitz & Associates, Inc., Portland, Oregon
JOHN H. ROBINSON, National Oceanic and Atmospheric Administration, Retired, Santa Barbara, California
NINA SANKOVITCH, Natural Resources Defense Council, New York, New York
ROGER E. VAN DUZER, Marine Operations, Shell Marine Department, Shell Oil Company, Houston, Texas
JOHN A. WITTE, Donjon Marine Co., Inc., Hillside, New Jersey

LIAISONS

KEN KEANE, Marine Environmental Protection Division, Office of Marine Safety, Security and Environmental Protection, United States Coast Guard, Washington, D.C.
WILLIAM PECK, Salvage and Diving, Naval Sea Systems Command, U.S. Navy, Arlington, Virginia

STAFF

CHARLES A. BOOKMAN, Director
RICHARD WILLIS, Consultant
AURORE BLECK, Administrative Assistant

MARINE BOARD

PREFACE

BACKGROUND

A major concern in promoting safe maritime commerce is the capability to effectively respond to marine casualties that threaten the environmental and economic resources of U.S. waterways and coastlines. Preventing marine accidents should be a perpetual goal of shipping interests and organizations, but the ability to react adequately after an accident to save all or part of an imperiled ship or cargo is an important element in preventing any environmental or economic harm. That capability—to render services to save maritime property, defined as "salvage"—is the focus of this study.

The National Research Council (NRC) addressed the U.S. capability to respond to an imperiled ship in 1982 and subsequently published a report entitled *Marine Salvage in the United States* (NRC, 1982). The study found, among other things, that the traditional salvage company, with dedicated vessels and personnel, was disappearing from the commercial salvage market due to high maintenance costs and fewer marine accidents. This trend has continued into the 1990s, along with a corresponding increase in the occasional use of general marine contractors for salvage. In addition, significant changes and pressures have emerged in both the environmental and the regulatory arenas during the past decade, heightening concern over the readiness in the United States to respond to a stricken ship and save the ship and all or part of its cargo.

SCOPE OF STUDY

The Committee on Marine Salvage Issues was established in April 1992, at the request of the U.S. Navy Supervisor of Salvage, to examine the issue of jettisoning cargo from a stranded vessel in order to lighten the ship and save it, thus averting further pollution. Subsequently, in response to a request from the U.S. Coast Guard, the committee's charter was expanded to include updating the 1982 Marine Board report (NRC, 1982), and in so doing to evaluate and make recommendations concerning national salvage capability.

The committee included shipping and salvage experts, legal and environmental protection specialists, and scientists and engineers (biographies of the members appear in Appendix A). Care was taken to ensure a balance of experience with the multiple interests related to this issue, including environmental, economic, and regulatory aspects. The committee was assisted by liaison representatives of the sponsoring federal agencies, the Navy, and the Coast Guard. The principle guiding the committee, consistent with NRC policy, was not to exclude any information, however biased, that might accompany input vital to the study, but to seek balance and fair treatment.

To fulfill its mission, the committee held a one-day symposium on the issue of jettisoning cargo from a stranded vessel in Washington, D.C., on February 23, 1993. Invited participants represented the U.S. Congress, federal and state agencies, and the fields of marine environmental protection, salvage, vessel operations, admiralty law, and marine insurance. A report based on the symposium (see Appendix B) has been published separately (NRC, 1994).

To carry out the second part of its mission, the reassessment of U.S. salvage capability, the committee undertook a number of information-gathering tasks. First, using updated accident scenarios from the 1982 report, a questionnaire was developed and distributed to solicit input on the status of salvage response efforts from salvage specialists and responsible state and federal agencies (respondents are listed in Appendix C). The committee analyzed the information it received and identified key issues for further review. Second, the committee held regional meetings to evaluate the salvage capabilities on the East, West, and Gulf coasts. Individuals with specific expertise were invited to focus on specific technologies and regional issues (participants in the meetings of the Committee on Marine Salvage Issues are listed in Appendix D). The committee also examined what the salvage industry's response would be today to the accident scenarios in the 1982 report. Finally, the committee evaluated all the information acquired, including its assessments of more than a dozen key issues and materials from the jettisoning symposium, to determine current U.S. salvage readiness and capability. A quantitative analysis of salvage equipment and personnel resources was beyond the committee's means. The review of the salvage posture was limited to activities on the East, Gulf, and West coasts, due to time and cost constraints and the determination that changes in those areas would be representative of other areas of the country. This limitation notwithstanding, some salvage incidents examined by the committee occurred in the Great Lakes, and some of the U.S. professional salvage capability assessed in the study is located in this region.

ORGANIZATION OF THE REPORT

This final report synthesizes all the information gathered by the committee:

- Chapter 1 reviews the importance of salvage in preventing marine pollution and in maintaining ports and waterways.
- Chapter 2 outlines the major changes that have occurred in the industry since the 1982 report and assesses regional salvage readiness and capability.
- Chapter 3 examines major issues facing the industry and alternative approaches for addressing the various concerns.
- Chapter 4 presents the committee's major conclusions and recommendations.

The report is intended to assist federal and state governments and the private sector in determining the policy, regulatory, and economic actions needed to ensure that the United States can respond effectively to marine accidents and prevent or minimize environmental and economic damage. It is hoped that the report also will help educate the public, the media, and all who are involved in any aspects of marine salvage.

ACKNOWLEDGMENTS

The committee greatly appreciates the valuable input and insight provided by the participants in the symposium and the regional meetings and by the respondents to the questionnaire. Jerry Aspland, Arco Marine, Inc. and a member of the Marine Board, served as the Marine Board's liaison with the committee. The committee also wishes to acknowledge the valuable input of reviewers, who did an exceptional job of providing specific suggestions and revisions; of the liaisons from federal agencies, who provided extensive information and support; and of Marine Board staff members, who provided unwavering support.

CONTENTS

Tables and Figures

Executive Summary

A key challenge of the 1990s for the marine industry—and for all who have an interest in marine transportation and protection of the marine environment— is to sustain a level of salvage capability that ensures effective response to marine casualties posing a risk to the environment and to waterborne commerce. The marine and salvage industries and governments must ensure that this capability exists and is sustained over time.

Salvage operations can help prevent pollution by providing assistance to a damaged or stressed vessel. For example, salvors can transfer cargo or fuel from a damaged vessel to a sound one (a procedure known as lightering), thereby reducing the threat of a spill. They can tow a damaged vessel to a safe harbor for repairs, or perhaps even repair the damaged vessel on site. They can jettison a portion of the cargo in order to minimize pollution or other damage. Salvors also can assist in minimizing or even preventing the closure of ports and waterways by rapidly repairing or removing vessels that are blocking passage.

The international marine salvage industry has undergone significant changes in the past two decades. Of most concern are the dwindling numbers of trained salvors and the decline in dedicated salvage resources worldwide. The general approach to salvage in the United States, too, is in a state of flux. Whereas historically the salvor directed the salvage response effort, management of vessel salvage response now is becoming a joint effort:

- The vessel owner is assuming a more active role in salvage response management, including, in some instances, contracting directly for specialized needs rather than relying on a single salvage company for all salvage services.
- The federal government, acting through the U.S. Coast Guard, has become proactive in handling vessel casualties that involve actual or threatened pollution, providing oversight, direction, and often active participation in the response effort.
- The salvor often is relegated to a consulting position, without direct input into the decision-making process.

Major changes have been made in the operational procedures and regulatory regime of the marine transportation industry. The traditional principles of salvage law were modified in the International Salvage Convention of 1989 to recognize the salvor's duty to protect the environment and to authorize a special compensation award to promote that duty. In the United States, environmental and coastal concerns led to passage of the Oil Pollution Act of 1990 (OPA 90, P.L. 101-380) and related state statutes. OPA 90 imposes new and expanded responsibilities on the oil-carrying segment of the marine transportation industry to prevent, plan for, and respond to oil spills.

REGIONAL ASSESSMENTS

In its assessment of salvage resources on the East, Gulf, and West coasts of the United States, the Committee on Marine Salvage Issues found that, in general, the nation's salvage capability and readiness appears to have increased in the decade since 1982. This readiness has accelerated, no doubt, as a consequence of OPA 90 and related state and industry actions. The improvement has not taken the form of a resurgence of the traditional salvage industry, with large salvage vessels and permanent stores of equipment and cadres of trained personnel. Rather, the increased capability stems mainly from the availability of fly-away[1] and prepositioned stockpiles of equipment, the development of networks of salvage equipment and expertise, and a general increase in the power of tugs used in coastwise and harbor towing. Also of note is the aggressive role the Coast Guard now assumes to ensure timely response to major accidents.

Resources available now that were not available in the 1980s include strategically deployed response assets provided by oil spill responders. These assets, while not designed to perform salvage, can provide assistance quickly due to their prepositioning near major shipping areas.

The level of salvage activity in the United States continues to be insufficient to support traditional salvage practices. As a result, salvage has become a secondary business for salvors and other marine contractors. Financial arrangements, such as retainers ensuring the availability of salvage resources required under the vessel response plan regulations arising from OPA 90, have improved the financial status of some salvage providers but do not offer enough incentive for companies to maintain dedicated vessels and crews awaiting offshore casualties. Nor is the practice of requiring retainers to ensure availability of response capability uniformly practiced among salvors. What is more, companies that can afford such investments would be reluctant to maintain often-unemployed assets producing limited return. In other words, with marine casualties in the United States occurring at a historic low rate, there is not enough salvage activity to make salvage a paying proposition for companies dedicated solely to salvage. Nor is the level of activity sufficient to attract and train future salvors.

In sum, although the salvage industry in the United States appears to be on an upswing—at least temporarily—as a result of OPA 90, the long-term prognosis may still be bleak unless ways can be found to train salvors and stimulate salvage-related business activity.

KEY ISSUES, CONCLUSIONS, AND RECOMMENDATIONS

Effective, time-critical salvage provides an important safeguard against environmental damage and commercial loss. It can prevent breakdowns, navigation errors, and other vessel mishaps from developing into more serious incidents, which could result in major pollution or channel and port closures. The salvage industry continues to face immense economic pressures. In general, the incidence of maritime casualties, and thus the need for salvage services, has declined by one-third since 1973. This trend is favorable for safety and environmental protection, but it has a negative effect on business conditions in the marine salvage industry.

Additionally, the nature of salvage has changed in the last decade. At least four major trends have been observed:

1. The motivation for maintaining a salvage capability has shifted from a private concern—protecting the vessel and its cargo—to a more public or societal interest in protecting the environment and economy from impacts of a vessel casualty.
2. The traditional dedicated salvage company with fully integrated capabilities has had to supplement its income by utilizing its salvage assets and

[1] Fly-away packages consist of specialized equipment, such as firefighting gear, diving systems, or pumps. This equipment is packaged and designed to be shipped on an airplane from storage locations to required destinations.

experienced personnel in more conventional marine business. Salvage has become only one segment of the organization's business.

3. Compensation for salvage has not kept pace with either the changing roles or increased financial risks of the salvor.
4. In the United States, the salvor no longer is in charge of the decision-making process employed in responding to a marine casualty, particularly one involving pollution or the threat of it. Instead, the salvor's role is to assist and provide direction to a Unified Command System involving the federal government, state government(s), and the responsible party (vessel or cargo owner or designee).

As a result of these trends, today the U.S. salvage capability is found in a small number of professional salvage companies, some dealing only with a specialized and limited aspect of salvage. Nevertheless, response to vessel casualties that have occurred in the United States since 1982 has been timely and effective. The committee did not identify a pattern of failure to respond because of the lack of salvage capability.

NATIONAL SALVAGE POSTURE

Despite the good response record of salvors in recent years, the committee perceives possible deficiencies in U.S. salvage capability. Conclusions regarding these potential deficiencies and recommendations for addressing them are outlined in the following sections.

COMPENSATION FOR SALVAGE

The committee concludes that, because of the change in societal values as they relate to salvage— from an emphasis on preventing or mitigating the loss of vessels and cargo to concern for preventing or mitigating environmental damage and other indirect consequences—the economic basis for rendering salvage services needs to be restructured. The committee therefore recommends:

The criteria for determining compensation for salvage in the United States should be updated to reflect changes in the business structure of the salvage industry and in societal values with respect to salvage and to incorporate the criteria delineated in the Salvage Convention of 1989.

The Salvage Committee of the Maritime Law Association should promulgate an updated regime of criteria for salvage awards, reflecting current salvage conditions, to be brought to the attention of courts and arbitrators without waiting for a case to be litigated or arbitrated.

NATIONAL SALVAGE POLICY

The national salvage policy continues to be that stated in the Salvage Facilities Act of 1948: that the public interest is served by maintaining salvage capability to provide for the national defense. In contrast to the fifty-year-old statement of national policy, the committee concludes that the primary public motivation for maintaining a salvage capability has shifted from national defense to include protecting the environment and the economy from impacts of a vessel casualty and that salvage fulfills additional functions in preventing or minimizing marine pollution, providing for public safety, and minimizing the disruption of port activities. The committee therefore recommends:

The Congress should update the national salvage policy to ensure that an adequate level of salvage capability is present in U.S. waters. The policy should clearly delineate the following as goals: to protect national security, to minimize or prevent environmental impacts due to pollution from marine casualties, to protect public safety, and to ensure minimal disruption to the U.S. economy resulting from marine casualties in the nation's ports and waterways.

The updated national salvage policy should specifically address the role of government agencies and their relationship to the private sector. At the very least, the policy should clarify the current roles of the U.S. Coast Guard and the U.S. Navy, as follows:

- **The U.S. Coast Guard should oversee or direct response to marine casualties in which there is peril to life, the environment, or other public interests.**

- **The U.S. Navy should advance salvage technology and practice through technical development programs; provide technical assistance in nongovernmental marine casualties requiring salvage; maintain contractual arrangements with private contractors for salvage services in excess of internal capability and make such services available on request; and train U.S. Navy and civilian personnel in salvage.**

The above government functions are essential to ensure that adequate salvage capability exists in the United States, and that there is adequate salvage response to marine casualties. Implementing this recommendation would, in the main, provide authority for current practice.

The national salvage policy should continue to clearly state that government assets will be used only when commercial assets are not available.

The 1982 National Research Council study on salvage recommended that the salvage issue be revisited in ten years. This study coincides with that time frame and finds significant changes requiring action that justify the 1982 recommendation. The committee therefore recommends that the national salvage posture be assessed again ten years hence.

SALVAGE ASSETS AND SERVICES

The committee deems it to be in the national interest to ensure that the U.S. salvage capability remain (at a minimum) at its current level. However, the committee's analysis of marine risks, casualty rates, and salvage business conditions supports the conclusion that the current level of capability cannot be sustained, nor identified gaps filled, without attention to acquiring, maintaining, sustaining, and positioning salvage assets, and training and retaining salvage personnel. Furthermore, the committee concludes that these measures require the direct assistance of both the marine transportation industry and government. The committee therefore recommends:

The U.S. Coast Guard, in consultation with the U.S. Navy Supervisor of Salvage, should develop and promulgate a list of active salvors in accordance with the general criteria proposed by the committee. The list could be used to prequalify salvage companies for inclusion in vessel owners' contingency plans for casualty response and may include salvors that specialize in specific aspects of salvage. As a matter of corporate

policy, companies needing salvage services should endeavor to contract for salvage services with professional salvage companies.

As new offshore tugs and escort vessels are built, designers and owners should be encouraged to add features to expand the response capability of those vessels to include some aspects of salvage. At a minimum, firefighting capability should be installed, as well as basic salvage equipment lockers.

Owners of all commercial vessels should be required to demonstrate that they have considered and are prepared to address, by contract or other approved means, key aspects of salvage capability as it relates to the safe conduct of their vessels in U.S. waters. (At present, this requirement applies only to vessels carrying petroleum as cargo.)

TUG AVAILABILITY

The committee concludes that geographic gaps may exist in the availability of adequate towing vessels for control and management of stricken ships, particularly in the Florida Straits and in areas of the West Coast. In light of physical and economic limitations to providing dedicated rescue towing salvage vessels, the committee recommends:

The U.S. Coast Guard should expeditiously complete studies mandated by OPA 90 on vessel routing schemes and exclusionary zones and vigorously pursue the institution, as appropriate, of the conclusions arising from those studies in both national and international forums.

SALVAGE READINESS OF VESSEL AND CREW

The voluntary nature of compliance with "salvage-friendly" vessel designs, such as that proposed by IMO Resolution A.535(13), has not had the expected result of widespread adoption of such features and the potential increase in the ability to save vessels in distress. The committee therefore recommends:

The Coast Guard should urge the International Maritime Organization to proceed with the revision of Resolution A.535(13), "Emergency Towing Requirement for Tankers," as expeditiously as possible and to promulgate the revised resolution in the form of a regulation requiring compliance by new and existing vessels. The Coast Guard should request that the IMO consider requiring emergency towing arrangements on vessels other than tankers.

In addition, the Coast Guard should urge vessel owners to incorporate other "salvage friendly" modifications, such as cargo piping system modifications, and institute formal salvage training for vessel crews.

TRAINING

The committee concludes that, with few opportunities to practice the salvage profession, the next generation of salvors is not being trained or attracted to the industry. The committee therefore recommends:

The U.S. Navy training facilities at Panama City, Florida, should be made available to selected industry and/or federal and state agency personnel,

sponsored by professional salvage companies and the respective agencies. Such training could be offered on a total reimbursement basis or financed by the Salvage Facilities Act.

The salvage and marine transportation industries should establish salvage training at a maritime institution on a cooperative basis (as was done with maritime firefighting schools). In addition, they should cooperatively develop a career track program within and across their respective industries to attract individuals to the salvage industry.

MARINE FIREFIGHTING

The committee concludes that marine firefighting experience in port areas is lacking. There are significant differences in marine firefighting capabilities among ports. Of particular concern is the interface between local fire departments and the marine transportation system. The committee further concludes that the location of prepositioned marine firefighting systems relative to vessel traffic patterns is a concern in some geographic areas. Of special concern are ports and areas with high passenger vessel concentrations. The committee therefore recommends:

The area planning process within the National Contingency Plan should include a review of local and area firefighting and salvage readiness and capabilities.

RESPONSE TO HAZARDOUS CARGO

The committee concludes that capability for responding to casualties involving hazardous cargo is limited. The committee therefore recommends:

The Coast Guard or another government agency should undertake a study on response to marine casualties involving hazardous cargo and the role of the salvor in the response.

JETTISONING

The committee concludes that ambiguities in federal and state oil pollution laws have created uncertainty concerning liability for acts of jettison. The committee therefore recommends:

The National Contingency Plan should be modified to incorporate criteria that would authorize intentional jettison when necessary. Modifications should also clarify the federal on-scene coordinator's ability to authorize incidental discharges resulting from ongoing cleanup operations, such as decanting from skimmers and pumping of engine rooms, through the daily work plans approved by the Unified Command System where appropriate.[2]

[2] The jettison issue was addressed by the committee in a separate report (NRC, 1994), which appears, with more specific recommendations, in Appendix B.

SAFE HAVENS

The committee concludes that, in the absence of predesignated safe havens, the risk is increased of marine casualties having catastrophic outcomes with environmental consequences. The committee therefore recommends:

The Coast Guard should promulgate the process by which a "safe haven" is identified. To the extent possible, area plans should evaluate candidate sites for potential safe haven areas.

1

Introduction

A key challenge of the 1990s for the marine industry—and for all who have an interest in marine transportation and protection of the marine environment—is to sustain a level of salvage capability that ensures effective response to marine casualties posing a risk to the environment and to waterborne commerce. The marine and salvage industries and governments must ensure that this capability exists and is sustained over time.

Salvage in the United States in the 1990s is shaped by trends in environmental protection and related local, national, and international legislation, as well as by changes in the marine salvage business over the past 20 years. The traditional role of the salvor has been to save a distressed ship, her cargo, and sometimes the lives of the crew. This role has changed with growing public concern for the environment. Presently, particularly in those casualties involving a ship carrying a potentially polluting cargo, salving the ship and cargo has become secondary to preventing or minimizing environmental damage. In other cases, where an accident occurs in a busy waterway and leads to closure of a port, the economic pressure for continued port operation tends to supersede concern over the value of the ship and cargo. In addition, private demand for salvage services has fallen, due to a declining number of vessel accidents requiring salvage (NRC, 1991; Tecnitas, 1992) and a trend among ship operators to employ nondedicated equipment and personnel to respond to marine casualties.

The shipping world—including shipowners, operators, cargo owners, and underwriters—uses the term "salvage" to describe all services rendered to save property from marine peril. This broad definition encompasses not only actions undertaken to save the vessel or cargo, but also wreck removal, harbor clearance, and deep search and recovery. Salvage includes:

- Providing firefighting assistance.
- Refloating a vessel from a stranding.
- Offloading cargo or water to prevent foundering, or removing sound cargo from impending peril.
- Shoring, patching, and making any temporary repairs to correct structural, stability, or mechanical problems.
- Rescue towing of an incapacitated vessel to a safe haven.
- Preventing pollution.

Such salvage operations are often time-critical in that success depends on timely action by experienced personnel and organizations.

A 1982 National Research Council (NRC) report entitled *Marine Salvage in the United States* (NRC, 1982) identified trends that redefined the salvage industry in the 1980s. Those trends included increasing vessel size, mounting complexity in cargo shipping and handling operations, increased carriage of hazardous cargoes, growing public and political interest in marine pollution, rising liability for spills of oil and other hazardous cargoes, shrinking traditional salvage capability, and development of new technology—including fly-away salvage equipment packages.[1] All these trends have continued.

In 1992, the Salvage Working Group, an industry organization representing salvors, shipowners, underwriters, and other interested parties, commissioned the Tecnitas Division of Bureau Veritas to study global salvage capacity and current levels of demand for salvage services and to examine how the relationship between demand and resources may change in the years ahead (Tecnitas, 1992). The study concluded that international salvage resources are in serious decline. This decline is primarily due to:

- A drop in the number of marine casualties over the past decade, the effects of which were masked in the mid-1980s by the Iran-Iraq War.
- A falling level of remuneration for salvors, to the point where the return is inadequate to support dedicated salvage capability.
- Fierce competition for salvage work from nondedicated[2] resources, aggravated by the search for low-cost salvage services by shipowners, insurers, and protection and indemnity (P&I) clubs.

Growing environmental concerns and increased liability for environmental damages have stimulated interest among vessel owners in seeing successful salvage carried out, thereby preventing uncontrolled releases of cargoes of fuel oil, crude oil, or hazardous materials. Environmental concerns also have stimulated greater public interest in salvage operations and have prompted some countries to become more involved in ensuring available salvage capability. Over the past ten years, interest in controlling the public consequences of casualties increasingly has overshadowed private concern for saving hulls and cargoes. However, compensation for salvors' efforts to avert or minimize pollution is only beginning to be integrated into the payment structure.

The future need for salvage services is related to the future risk of vessel accidents. Several factors can be expected to affect the future accident rate. On the negative side, low freight rates, brought about by the downturn in the world economy and overtonnaged markets, have put shipowners under financial pressure, which can lead to a lowering of maintenance and crewing standards. Indeed, some owners have succumbed to this pressure. Compounding this problem is the aging of the world fleet, a trend that heightens the need for maintenance.

On the positive side, the problems have been recognized by governments and private industry, which have responded with initiatives designed to prevent a reversal of the trend of declining marine transportation system accident rates. The Oil Pollution Act of 1990 (OPA 90, P.L. 101-380) in the United States and Regulations 13F and 13G of MARPOL 73/78 Annex 1 from the International Maritime Organization (IMO) require the removal of older tank vessels from the world market on a fixed time schedule and major structural changes, such as double hulls, for replacement vessels. These laws also mandate enhanced maintenance and inspection procedures. In addition to legislative efforts, segments of the oil industry, the tanker industry, and the classification societies have taken steps to improve tank vessel quality. This is a

[1] Fly-away packages consist of specialized equipment, such as firefighting gear, diving systems, or pumps. This equipment is packaged and designed to be shipped on an airplane from storage locations to required destinations.

[2] Nondedicated resources are those used in applications other than marine salvage, but which can be substituted and/or utilized in a marine casualty response.

worldwide effort, but in the United States the liability regime and the resulting financial risk involved in using substandard vessels have led to a redoubling of efforts to improve marine safety.

U.S. initiatives have, to date, been directed at the tanker industry. Passenger and dry cargo ships, however, are at equal risk, and carry significant quantities of fuel oil and, at times, potentially polluting hazardous cargoes. All vessels carry crew, and the protection of life at sea is of primary importance. Vessel construction and operational procedures in these domains have yet to undergo the type of rigorous review that the tanker industry has undergone.

THE IMPORTANCE OF SALVAGE IN PREVENTING MARINE POLLUTION

Oil spills from vessels in distress can result in severe economic and environmental damage. Vessels that incur structural damage, are involved in collisions, or are stranded pose a risk of pollution due to loss of cargo or fuel oil. The discharge of potentially polluting cargoes such as oil, chemicals, and even otherwise nonhazardous materials, such as edible oils and other nutrients, can have adverse effects on the marine environment. These effects may vary widely from incident to incident. The nature of the substance spilled, the location of the accident, and weather conditions at the time are among the most important factors in determining the seriousness of the event (Engelhardt, 1994).

Among potentially polluting cargoes lost in shipping accidents, oil and its products are by far the most common. Over the past 20 years, extensive research has been directed at understanding the effects of oil spills and factors that govern the environmental consequences of a given incident (NRC, 1985). Certain refined products, such as diesel oil and gasoline, generally are more toxic than crude oils, on a volume-for-volume basis, in that they contain a higher concentration of lighter aromatic hydrocarbons. The aromatic components of oil pose the greatest potential of acute toxicity to both plants and animals and cause a variety of sublethal effects, including impairment of reproduction, physiological stress, and reduced growth rates. However, light products generally evaporate, dissolve, and otherwise disperse much more readily in the environment than do heavier oils, which have greater smothering potential and are more difficult to clean up.

A wide variety of chemical products other than oil are spilled every year in U.S. waters. Sulfuric acid and ammonia generally top the list because they are shipped in high volumes. Spills of these substances have been associated with major fish kills in localized areas. Less common, but more toxic and more persistent substances have caused prolonged restrictions on fishing and even lengthy closures of important waterways to ship traffic. Occasionally, spills of seemingly innocuous products, such as fertilizers, have upset the delicate nutrient balance in enclosed estuaries and bays, resulting in severe short-term damage.

Historically, spill response has not been very effective. Only 10 to 15 percent of oil typically is recovered following a spill (Office of Technology Assessment [OTA], 1990). Hazardous chemicals pose a particularly difficult challenge. The Chemical Manufacturers Association (CMA) recently concluded that the capability to respond to a chemical spill of any significant magnitude in the sea does not exist anywhere in the United States (CMA, 1992).

Salvage operations help prevent pollution by providing assistance to a damaged or stressed vessel. For example, salvors may be able to transfer cargo or fuel from a damaged vessel to a sound one (a procedure known as lightering), thereby reducing the threat of a spill from an unstable vessel, tow a damaged vessel to a safe harbor for repairs, or perhaps repair on site. The *Exxon Valdez* accident[3] is a case in point.

[3] All instances of salvage in U.S. waters that were reported to the committee are listed in Appendix E.

Salvors successfully removed more than 1.02 million of the 1.26 million barrels of North Slope crude oil that were aboard. Salvors also readied the damaged vessel to be towed to a repair yard and supervised the successful operation. There are a number of other recent examples of successful salvage actions where potential pollution of the environment was avoided or minimized:

- In 1993, a collision involving the tank barges *Ocean 255* and *BT 155* and the freighter *Balsa* resulted in the loss of about 6,000 barrels of refined product. The response effort was successfully carried out by the companies involved, their spill contractors, and the U.S. Coast Guard strike team. Two hundred thousand barrels of product were offloaded to other vessels without further pollution.
- In 1992, the tankers *Radwan* and *Argo Hebe* collided in the Straits of Malacca, and fires broke out on both vessels. The *Radwan* was carrying 178,000 barrels of gasoline, the *Argo Hebe* 2 million barrels of crude oil. The *USS Beaufort*, a U.S. Navy fleet tug, which happened to be a few miles away, witnessed the collision. Within 30 minutes, the *Beaufort* was alongside the tankers, and Navy salvors suppressed both fires and rescued 20 sailors. (It is unlikely that the *Radwan* would have survived without such prompt assistance.) No pollution occurred.
- In 1991, the *Kirki* lost its bow off the West Australian coast. The *Kirki* was carrying 525,000 barrels of oil. Salvors transferred all the cargo to another tanker and towed the *Kirki* to Singapore, averting any significant oil pollution.
- In 1990, fire broke out in the engine room of the *Mega Borg*, which was 60 miles offshore from Galveston, Texas, close to important marine habitats. A promptly launched professional salvage effort extinguished the blaze over several days. The 120,000 barrels of oil that were lost were consumed almost entirely by the fire; the remaining 857,000 barrels were saved. The National Oceanic and Atmospheric Administration (NOAA) reported that no significant ecological damage resulted from the spill.
- In 1989, the tanker *Phillips Oklahoma* collided with another vessel and burned off the eastern coast of England. The tanker was carrying close to 357,000 barrels of crude oil; over 5,000 barrels spilled following the collision. Due to timely salvage assistance, the fire was extinguished, and all remaining cargo was transferred to another vessel.
- In 1989, the tanker *Pacificos* incurred serious structural damage off the South African coast while carrying over 1.67 million barrels of oil. The casualty was towed to the Mozambique Channel. Despite an initial loss of over 50,000 barrels, all remaining cargo was transferred to another tanker without further pollution.

The extent to which salvage can prevent or minimize pollution depends on the availability of salvage resources. Recognizing this dependency, some nations now provide such resources. For example, the governments of Spain and South Africa have retained salvage tugs to assist in response to ships in distress along their coasts. In France, a joint venture between the government and private salvors provides protection for the English Channel and the Atlantic and Mediterranean coastlines of France and Corsica. Since 1979, the retained salvage tugs have responded to more than 430 distress alerts and have carried out nearly 80 salvage operations. Regrettably, the French government has reduced the number of subsidized vessels on alert status. The objective of these government-initiated partnerships is to ensure that the capability is in place to respond rapidly and effectively to all marine emergencies, especially those with the potential to pollute the environment. The United States has not made a similar commitment but does enter into contractual arrangements with commercial salvors to maintain salvage readiness to meet the Navy's needs (see Appendix F).

THE IMPORTANCE OF SALVAGE IN MAINTAINING PORTS AND WATERWAYS

Following a collision, stranding, or sinking or an oil spill, a port or channel may be partially or completely closed, sometimes for days or even weeks. The costs of such closure, including lost business and maritime traffic disruption, can run to many millions of dollars. The collision and subsequent sinking of the sulfur barge *Duval 2* in the Houston Ship Channel in 1992 is a case in point. The barge was split almost in two and effectively blocked the channel to all traffic for 24 hours. Limited one-way traffic resumed after the barge was pushed to one side of the channel; two-way traffic was not restored for almost two weeks. Some 200 ships were delayed, causing operational disruptions to local refineries and other industries. The added operational expenses per vessel, for each day delayed, averaged $20,000 (American Waterways Operators, personal communication, April 1993).

Salvors assist in minimizing or even preventing the closure of ports and waterways by rapidly repairing or removing vessels that block passage, or by minimizing the amount of oil spilled from damaged vessels and thus the duration and extent of cleanup operations. In the *Duval 2* incident, salvage was prolonged because resources were slow in arriving. Salvage equipment had to be dispatched from New Orleans, and passage to the job site was delayed by fog. A foreign-flag heavy lift barge was available in the Houston area, but requests for its use were not made.

The closure of ports and waterways is a particular problem in the Gulf of Mexico region, due to the narrow channels that must be negotiated to reach or depart from key ports and heavy traffic in those channels. The problem of port closure due to vessel incidents, however, is not limited to Gulf ports. For example, in 1990, the oil tanker *BT Nautilus* grounded in the Kill van Kull Channel of New York Harbor, spilling over 6,200 barrels of heavy oil and causing partial closure of the busy waterway for several days.

2

CHANGES IN THE MARINE SALVAGE INDUSTRY SINCE 1982

THE BUSINESS ENVIRONMENT

The international marine salvage industry has undergone significant changes in the past two decades. Dwindling numbers of trained salvors and declining dedicated salvage resources worldwide cause the most concern. The demand for long-distance international tows, once a primary alternative source of income for the salvage industry, also has dropped dramatically with the development of specialized heavy-lift vessels and changes in world trade.

While the need for traditional salvage services has declined significantly—particularly dedicated, on-call salvage tugs—alternative uses of commercial equipment for salvage have increased. High-horsepower tugs and supply vessels normally employed in offshore oil fields have been used for salvage and firefighting. Conventional tugs, accompanied by a specialized salvage vessel, successfully towed the *Exxon Valdez*[1] from Alaska to California. The *Mega Borg* was salved through the use of specially equipped tugs and supply vessels normally engaged in the offshore oil and gas industry. It is important to note, however, that salvage operations in both incidents were directed by experienced salvage masters.

The general approach to salvage in the United States is in a state of flux. Historically, salvors have directed salvage response efforts, but now vessel casualty salvage response is likely to be jointly managed. Vessel owners are assuming a greater role in salvage response management, including in some instances contracting directly for specialized salvage needs, rather than relying on a single salvage company. This evolution has been accompanied by the development of service industries that provide specialized salvage equipment and services, such as pumps and fenders for cargo transfer and lightering and marine firefighting. The federal government, acting through the U.S. Coast Guard, has become proactive in handling vessel casualties that involve actual or threatened pollution. The Coast Guard provides oversight and direction and often actively participates in response efforts. The salvor often is relegated to a consulting position, with no direct input into decision making.

The salvage industry has also been affected by advances in technology. The development of portable, fly-away salvage-related systems, such as those in the areas of firefighting and lightering, has had significant impact on the structure of the salvage industry. Worldwide, single-lift crane capacity now exceeds 14,000 tons. Salvage recovery using remotely operated vehicles (ROVs) has been demonstrated successfully by the U.S. Navy in water as deep as 20,000 feet. Commercial recovery using ROVs is now done routinely in water over one mile deep. Diving, especially deep saturation diving, is safer and more productive than ever in salvage situations. Saturation dives at depths of 1,000 feet are not uncommon. Advanced information technologies, such as stability programs that can be run on portable computers and

[1] All instances of salvage in U.S. waters reported to the committee are listed in Appendix E.

remote sensing techniques, provide salvage personnel with information that can improve decision making. Economic and incident response decision-making constraints, not technology, are the limiting factors in salvage today.

With the high capital cost of dedicated salvage equipment, the use of fly-away salvage units is rising in the United States. These are equipment packages that can be transported by commercial air freight and typically contain pumping, firefighting, and basic salvage tools and spill response equipment. The fly-away concept involves rapid deployment of response equipment and personnel within hours of a casualty. Although the concept has been well publicized, to fully realize the benefits of fly-away capability the physical presence of significant seagoing assets also is needed.

The government continues to maintain a salvage response capability, but these resources are limited—usually from the U.S. Navy Supervisor of Salvage. They are made available for commercial use only when commercial assets are not available. The Navy Office of the Supervisor of Salvage still relies heavily on its commercial salvage contractors (three at this writing) to carry out actual operations and to provide salvage engineering and related technical assistance beyond the Navy's own capability. The Coast Guard, by expanding its strike team concept, has increased the number of people and equipment capacity available to respond to spill and salvage incidents. The Coast Guard's equipment and procedures closely parallel that of salvor and/or oil spill response organizations, although it has not been directly tasked to include salvage response as one of its missions. Recently, both the Navy and the Coast Guard have been heavily involved in salvage situations, particularly those involving extensive pollution or the threat of it. Traditionally, however, the government has participated in commercial salvage only as a salvor of last resort.

Two additional factors have influenced the marine salvage industry since 1982. First, those responding to a salvage situation historically were concerned with saving the hull and cargo, and salvage decisions were made solely by the salvor and shipowner. Today, when pollution is involved, the response is highly regulated, and the outcome is determined by a consensus-based decision-making process involving the salvor, shipowner, and federal and state governments. Second, exceedingly high liability and cleanup costs of an environmental casualty have underscored the importance of adequate salvage response in preventing severe economic repercussions to the owner and underwriter. Significantly, tank vessel contingency plans required by the Oil Pollution Act of 1990 (OPA 90) require vessel operators to certify that adequate salvage capability is available, through either contract or other approved means (33 CFR 155.1035).

THE POLITICAL ENVIRONMENT

Over the past 20 years, significant changes in the maritime industry and the world at large have prompted alterations in long-standing principles of salvage. The primary force driving these shifts has been increasing interest in environmental protection. This interest was sparked by major oil spills, beginning in 1967 with the loss of the *Torrey Canyon* off the coast of England, and continuing with the 1978 *Amoco Cadiz* spill near France, the 1989 *Exxon Valdez* spill in Alaska, and more recently the *Aegean Sea* spill in 1992 off Spain and the 1993 *Braer* spill in Scotland.

Continuing pressure from environmental concerns and coastal interests resulted in major changes in the regulatory regime of the marine transportation industry since 1982. The traditional principles of salvage law were modified in the International Salvage Convention (ISC) of 1989. This convention, ratified by the United States but not yet in force internationally, specifically identifies the salvor's duty to protect the environment and authorizes a special compensation award to promote it. Prior to these modifications, the ISC of 1910 provided that the salvor's award would depend on the voluntary nature, difficulty, and especially the success of the salvage efforts, salvage compensation was awarded solely for success in saving vessels or cargo in distress by those under no legal obligation to do so. If the cargo or vessel were lost despite the best efforts of the salvor, no award was made—thus, the adage "no cure, no pay," which was reflected in contracts for salvage services such as Lloyd's Open Form (LOF).

The LOF was modified in 1980 following the *Amoco Cadiz* casualty to provide for reimbursement of expenses incurred by salvors to minimize or prevent pollution, even if the salvor was eligible for no other award. In 1990, reflecting the Salvage Convention of 1989, the LOF was again modified to reward salvors' efforts to prevent or minimize pollution. The salvor's skill and efforts in preventing and minimizing pollution were added to the criteria for fixing the reward. Additionally, the 1990 changes provided that the salvor was entitled to special compensation, up to 130 percent of the salvor's expenses (200 percent in special circumstances) for efforts to prevent or minimize pollution in the event that the salvor's award otherwise would be less than the special compensation. An award for a salvage operation undertaken under an LOF 90 has already been granted. This was in the case of the 1991 grounding of the *Yinka Folawayo* off the coast of Spain. Salvors were paid under the contract for their salvage *and* pollution prevention efforts; compensation was granted despite the total loss of the vessel.

Although U.S. salvage companies tend to be more willing than foreign salvors to work under terms other than the LOF, the philosophy that produced the changes in the LOF will have an effect on all future salvage contracts, even in the United States.

In the United States, environmental and coastal concerns led to passage of OPA 90 and related state statutes. OPA 90 imposes new and expanded responsibilities on the oil-carrying vessels of the marine transportation industry to prevent, plan for, and respond to oil spills. Regulations stemming from the act also require oil-carrying vessels to make response plans that identify and plan for salvage resources. In addition, port-area contingency plans must identify and plan for deployment of salvage resources. Thus, OPA 90 fosters increased awareness of the role of salvage in pollution prevention and mitigation, presumably a positive influence on the future of salvage resources in the United States. The terms "salvor" and "salvage resources" in the regulations have yet to be defined.

CHANGES SINCE THE 1982 REPORT

The 1982 NRC report, *Marine Salvage in the United States*, contains various conclusions and recommendations. Those recommendations and related changes enacted to date are shown in Table 2-1.

1982 REGIONAL SALVAGE ASSESSMENT

Conclusions in the 1982 NRC report, *Marine Salvage in the United States*, reflected various regional concerns about salvage capability. These concerns are summarized below.

East Coast

- Major salvage incidents in this region were too infrequent to justify maintenance of a dedicated commercial salvage ship.
- Salvage was expected to continue to involve improvised assemblies of assets on an incident-by-incident basis.
- Actions taken to stabilize distressed vessels, such as flooding control and dewatering (removing water from cargo tanks), were in regular use using state-of-the-art equipment.

TABLE 2-1 Recommendations and Conclusions from the 1982 NRC Report and Actions to Date

RECOMMENDATIONS	ACTION
Arbitrators and the courts should make more generous salvage awards. Companies that maintain reasonable salvage competence and readiness as demonstrated by personnel, equipment, and management should receive enhanced awards.	No trend is evident in the United States because few salvage cases have been taken to arbitration or have been litigated. The only cases that must be decided by U.S. courts are those in which the U.S. government is involved. A negotiated or day rate is normally used in the United States in salvage situations, rather than LOF or other arbitration. There is some indication that awards are increasing and salvage efforts to prevent or minimize pollution are being compensated.
The Coast Guard and other agencies should refrain from undertaking salvage operations that can be accomplished by the private sector.	Government policy continues to rely on the private sector for salvage response. However, the Coast Guard's interpretation of OPA 90 is that it requires the agency to be very active in any salvage situation that involves potential pollution. Coast Guard and Department of Defense (DOD) assets can be expected to be accessed in any major incident.
The Coast Guard should develop criteria for safe havens (ports of refuge for damaged vessels).	With the exception of Prince William Sound, Alaska, no action has been taken on this recommendation.
Public laws should be amended to absolve salvors from civil penalties for pollution that occurs as the result of prudent professional salvage activities.	No action has been taken.
The United States should ratify the International Convention on Civil Liability for Oil Pollution Damage, 1969, and the International Convention on the Establishment of an International Fund for Compensation for Oil Pollution Damage, 1971.	No action has been taken.
In those instances where salvage will contribute to environmental or public safety, but private salvage efforts are not timely, the Coast Guard should oversee salvage operations, using private sector assets as available.	This is now Coast Guard policy.
The U.S. Maritime Administration and the Navy should investigate the commercial feasibility of European-design tugs, with or without government support, and should develop designs as appropriate.	No action has been taken.
Companies engaged in salvage should improve the planning and development of salvage systems to ensure adequate salvage response.	Low level of salvage activity has limited investment in improving capability. No trend was discernible until passage of OPA 90. Industry response to OPA 90 requirements appears to be providing some funds and initiative for improving capabilities.

Table 2-1 continued

RECOMMENDATIONS	**ACTIONS**
Ship operators should develop corporate contingency plans for marine emergencies that include specific plans for salvage.	Contingency planning efforts have been extensive throughout the tanker industry, with an integral emphasis on salvage. No trend was discernible until passage of OPA 90. OPA 90 regulations require salvage planning for tank vessels only.
The National Contingency Plan (NCP) should be amended to address salvage.	Salvage per se has not been addressed in the NCP. As a result of OPA 90, salvage may be addressed specifically in the revisions to the NCP.
The Coast Guard should establish a system to gather, store, and provide environmental information in support of pollution-response planning that includes potential salvage operations.	The Coast Guard has initiated various efforts to provide such information. Salvage assets have not been a priority in these efforts.
The Navy and Coast Guard should encourage ship and terminal operators to plan and prepare for salvage operations.	No formal activity was initiated prior to OPA 90.
The Congress should update the statement of national salvage policy (10 USC 7361-7367).	No action has been taken.
The Navy should continue to oversee the national salvage posture and should periodically audit and report on U.S. salvage capability.	The Navy has addressed its own needs in a comprehensive study entitled *Salvage 2010*. Commercial salvage capability is the subject of the present NRC study, commissioned by the Coast Guard and the Navy. The Navy recently formed a Salvage Executive Steering Committee to examine current DOD marine salvage needs and to chart future Navy salvage force levels.
With assistance of the Coast Guard, the Maritime Administration, and U.S. industry, the Navy should continue to train the next generation of (salvage personnel) and develop, test, and stockpile improved salvage systems, equipment, and components.	The Navy continues to be the leader in salvage training outside of the private sector. The Navy, like the private sector, is faced with limited training opportunities. Billets for salvage experience are decreasing with declining demand. The Navy continues to develop and maintain state-of-the-art equipment for salvage.
The Navy should establish a subcommittee to encourage improvements in the national salvage posture.	No action has been taken. The Navy does issue contracts for salvage support activities that have the effect of "encouraging improvement in national salvage posture."

- Firefighting technology, systems, and level of training were inadequate for handling most shipboard fires, and there were few specialized firefighting systems or crews.
- No offshore firefighting vessels were available.
- No designated safe havens were available.
- No commercial vessels designed or operated specifically for complex salvage operations were available.
- A significant number of vessels capable of rescue towing were available on the East Coast, although none was built or crewed specifically for offshore salvage, and most had design constraints that limited salvage performance.

Gulf Coast

- Marine casualties were too infrequent to sustain commercial interest in salvage.
- Many vessels and considerable specialized marine equipment were available, due to extensive offshore oil and gas development.
- There was a scarcity of dedicated or purpose-built rescue tugs, shoreside stores of salvage gear, and specialized craft such as pulling vessels, heavy-lift vessels, or lightering barges. General-purpose vessels and gear were abundant throughout the Gulf.
- Although firefighting technology was more advanced in the Gulf than elsewhere, several improvements were needed to improve response to shipboard fires. Equipment was not standardized, most vessels were not equipped with up-to-date technology, and personnel were not trained in shipboard firefighting techniques.

Pacific Coast

- Physical capabilities and salvage assets, except those needed to fight fires, generally were adequate. Also needed were trained firefighters and technology for dealing with hazardous cargoes.
- Rescue towing coverage generally was adequate, although there were deficiencies in time-critical response in certain areas, such as remote Alaska.
- Safe havens existed, but their availability was questionable due to the difficulty of obtaining approval to enter these areas.

CURRENT ASSESSMENTS OF REGIONAL SALVAGE READINESS AND SALVAGE CAPABILITY

The committee used three types of information sources to assess salvage readiness and capability in the three regions of the United States. First, a comprehensive questionnaire was developed and sent to the various segments of the salvage industry, as well as to individuals, organizations, and government agencies with an interest in the industry. The results of the questionnaire (respondents are listed in Appendix C) were reviewed by the committee to ascertain areas of interest for future investigation. Second, regional assessment meetings were held on each coast (attendees are listed in Appendix D). These meetings featured invited presentations and open discussions; and issues identified in the review of the questionnaires, as well as additional issues that surfaced, were addressed. Third, the committee analyzed, through desk audits and tabletop exercises, what the salvage industry's response would be in today's environment to the same basic accident scenarios used in the 1982 report (NRC, 1982). (The scenarios are presented in Appendix G.) The results of all these efforts are summarized for the three regions in the following subsections.

East Coast

Salvage Capability

Salvage capability in several areas along the East Coast either has increased or is increasing. In 1982, no dedicated salvage vessel was present on the East Coast; since then, an offshore supply vessel has been acquired and modified for salvage and emergency response. The vessel is based in the New York region, and although available for other commercial work, is primarily tasked to support salvage efforts.

Firefighting resources were limited in 1982 to fire boats and to a few tugs equipped with fire monitors. The number of tugs with fire monitors has increased. Purpose-built firefighting equipment, prepackaged for air, truck, and water deployment, is maintained at various locations on the East Coast. Plans call for predesignated personnel to meet with the equipment at accident locations. Marine firefighting teams have been organized. The analysis of the salvage scenarios suggested that sufficient firefighting equipment could be brought to the scene of an incident anywhere on the East Coast and begin operation within 12 hours.

In 1982, many of the accident scenarios developed by the NRC committee relied on foreign salvage masters and salvage companies. Today, at least two U.S. companies provide full-service salvage response throughout the East Coast (and elsewhere) using company-owned and/or contractually provided, prepositioned response resources. U.S. divisions of the international salvage companies also are still available, although without assets on the East Coast.

Rescue Towing

Tugs designed, equipped, and crewed for ocean rescue towing continue to be scarce. Although the number of tugs with greater than 3,000 horsepower has increased from approximately 70 to about 140, they are not designed, equipped, or crewed for rescue towing. However, various towing companies remain willing to come to the rescue of a distressed vessel, and they expressed confidence in their ability to "control" such a vessel, thereby limiting its potential impact on shorelines, while waiting for other assets to arrive for long-range rescue towing. With the exception of the Florida Straits, sufficient equipment for control could be on-site within six hours in all of the exercises undertaken. The Florida Straits presents a special problem, in that the timing of assistance to a disabled ship would depend on the tug/barge traffic transiting those waters at the time of the incident. The extent of this traffic varies, but initial investigations suggested a response time of 12 hours.

Firefighting Capability

Salvage capability is increasing, with the availability of fly-away marine firefighting systems, including predeployed units on the East Coast and the owners inclusion of firefighting capability in vessel modifications and new buildings. However, the new level of resources has yet to be tested in real-life situations. Contingency planning for the various ports and areas has not yet incorporated the availability of fly-away marine firefighting systems and personnel, and local assets required for fast response needs of special situations such as passenger vessels are still limited.

Salvage Response Vessels

There continue to be significant numbers of tugs and barges on the East Coast to assist in a marine casualty, with availability dependent on their employment at the time of the casualty. Traditionally, these vessels have been supplied quickly for use in salvage response. Throughout the study, tug owners reported to the committee that they would provide vessels for salvage, releasing them from other duties where necessary and possible because it is in the response vessel owner's best interest, both economically and politically. Owners explained that they must serve the companies that normally employ their vessels; that they have a traditional responsibility to come to the aid of any distressed vessel; and that they have an interest in protecting the environment and the marine transportation system from disastrous incidents that may result in additional regulatory and economic burdens, such as more complex liability regimes and increased insurance costs.

Lightering Assets

Lightering activities are centered in Delaware Bay, and to a much lesser extent in New York Harbor. In these areas, lightering assets are available at very short notice. In other areas, resources are found at the various marine liquid cargo terminals. In the tabletop exercises undertaken by the committee, sufficient lightering assets were able to be marshalled at the casualty site within 12 hours.

Gulf Coast

Salvage Capability

The Gulf Coast area, from an overall perspective, has sufficient salvage resources and personnel to respond to a vessel casualty requiring salvage.

Rescue Towing

Rescue towing for very large crude carriers (VLCCs) is the biggest challenge in the Gulf of Mexico. This class of vessel trades normally within the area, utilizing the Louisiana Offshore Oil Port (LOOP) and the offshore lightering areas, and requires large tugs with horsepower in excess of 3,400 shaft horsepower (SHP) for towing and/or control in the event of a disabling incident with loss of power and/or steering. There remain a significant number of large, powerful tugs, including rig-moving and anchor-handling supply boats with horsepower of 4,200 to 13,000 in the Gulf. The rig-moving vessels (maximum horsepower of about 7,200 SHP) normally are equipped with towing winches, towing wire, and bridles. These vessels normally are employed on short charter contracts and are available for hire between jobs. The anchor-handling vessels have superior horsepower but are not ideal for rescue towing because of their design. Availability of these vessels is subject to international drilling activity.

In view of the availability of tugs capable of towing large vessels, in the committee's judgment, rescue towing capability in the Gulf is adequate.

Firefighting Capability

Firefighting assets, while more significant in the Gulf than on other U.S. coasts, still are limited. Fire monitors of greater than 5,000 gallons per minute (gpm), capable of providing assistance in an offshore fire, are found on approximately 14 vessels in the Gulf. These vessels are employed primarily in the offshore oil and gas industry, and their availability for salvage depends on other assignments. Significant fly-away firefighting equipment has been developed and deployed in the region, with foam stockpiled at various locations along the Gulf, including Florida, with a corresponding increase in the marine firefighting response capability. In general, the firefighting capability in the Gulf can be said to be better than in 1982. Numbers of experienced firefighters are limited but adequate for projected needs.

Salvage Response Vessels

There continue to be no dedicated salvage vessels in the U.S. Gulf. The salvage industry continues to rely on vessels of opportunity. The Gulf has significant vessel resources due to the offshore oil and gas industry that could provide emergency towing and other assistance for salvage response. The salvage effort for the *Mega*

Borg, for example, had six vessels on-site within 36 hours, with the capability of providing 67,500 gpm of firefighting water, at least two of which were capable of performing any rescue towing assistance needed.

Lightering Assets

Lightering assets—tankers, fenders, hoses, and support vessels—are readily available throughout the Gulf, and ship-to-ship offshore lightering is very common.

Channel Blockage

A major concern in the Gulf is channel blockage. From a salvage perspective, the key issue is availability of equipment to remove channel obstructions resulting from casualties. To a greater degree than on other coasts, the marine transportation system in the Gulf depends on the use of narrow ship channels. Casualties such as the *Duval 2* sulfur barge collision and sinking in the Houston Ship Channel can have a major effect on shipping in the affected port and consequently on the economy. Significant heavy-lift assets are available in the Gulf, but they are located in areas where they can be employed in the marine construction industry, not necessarily near the scene of a casualty. In the *Duval 2* incident, salvage assets deployed from New Orleans were delayed by fog and did not arrive on the scene for several days.

Pacific Coast

Salvage Capability

Salvage capability and readiness on the Pacific Coast continues to be generally adequate, with some exceptions. The presence of major towing and tug companies on the Pacific Coast continues to be a strong force in salvage readiness and response. The introduction of powerful tractor tugs equipped with firefighting monitors for tanker escort service in Puget Sound (also anticipated in California waters) has added significantly to the salvage capabilities and readiness on the Pacific Coast. This region also has the only dedicated salvage vessels in the United States.

Given the current level of salvage activity, there does not seem to be sufficient salvage work to keep the dedicated vessels gainfully employed (presently even these vessels rely on on-call crews). Without more activity or new revenue streams, the capacity level in this region may decline as one or more companies leave the field.

Rescue Towing

The availability of vessels capable of rescue towing is of less concern on the Pacific Coast than in other areas. This is mainly due to the accessibility of large commercial ocean-going tug fleets and the U.S. Navy presence in the mid-Pacific. However, there are significant gaps in coverage in remote areas of Alaska and in some areas of Northern California and Oregon, due to the long distances between ports.

Firefighting Capability

Firefighting capability remains the major concern in the Pacific region. Little traditional firefighting capability exists within the coastal port areas, and the interface between local authorities and the marine industry may not be adequate. The capability is increasing with the addition of required escort tugs for oil tankers in Puget Sound and those anticipated in California. These tugs are being equipped with fire monitors with over 5,000-gpm capability. The planned prepositioning of fly-away marine firefighting systems in the region will add significantly to the response capability. As in other areas, the added capabilities have not been incorporated in port and area contingency planning.

Salvage Response Vessels

The West Coast has the only dedicated salvage vessels in the United States. Located in Astoria, Oregon, and Seattle, Washington, these vessels offer excellent salvage work platforms and workshops. Due to the low level of salvage activity, the vessels have skeleton crews and rely on on-call personnel for on-scene salvage. The continued availability of these vessels, in the light of current economic conditions and their specialized use, is unlikely.

Lightering Assets

On the West Coast, with its deep-water ports, lightering is not a primary activity. The major towing and tug companies in the region provide significant resources to support lightering needs in the event of a casualty. The substantial tanker traffic on the coast adds to the availability of emergency storage and offloading capability.

Summary

In general, the salvage capability and readiness of the United States appears to have improved in the decade since 1982, accelerated as a result of OPA 90 and related state and industry actions. The improvement has not taken the form of a resurgence of the traditional salvage industry, with large salvage vessels, permanent stores of equipment, and a cadre of trained personnel. Rather, the increased capability stems mainly from the availability of fly-away and prepositioned stockpiles of equipment and the development of networks of salvage equipment and expertise. Also of note is the aggressive role the Coast Guard now assumes to ensure timely response to major accidents. As was evident in the response to the 1993 accident in Tampa Bay, Florida, the Coast Guard is not reticent about taking charge and participating actively in salvage when, in the opinion of agency officials, other action is not proceeding at a level adequate to address the risk involved.

Current resources, which were not available in the 1980s, include strategically deployed response assets provided by oil spill responders. These assets include dedicated response vessels and barges. Although they are not designed to perform salvage, they can provide assistance quickly due to their prepositioning near major shipping areas.

The level of salvage activity in the United States continues to be insufficient to support traditional salvage companies economically. As a result, salvage has become a secondary business for salvors and other marine contractors. The recent development of a system of retainers would improve the financial status of some salvage providers, but by itself may not provide enough incentive for companies to maintain dedicated vessels and crews awaiting offshore casualties. Moreover, companies that can afford such investments may be reluctant to have often-unemployed assets that produce only a limited return. In other words, with marine casualties in the United States occurring at a historic low rate, there is not enough salvage activity to make salvage a paying proposition for dedicated companies or to encourage investment in dedicated salvage equipment and facilities. Furthermore, the level of salvage activity is not sufficient to attract and train future salvors.

In sum, although the salvage industry in the United States appears to be on an upswing—at least temporarily and in part as a result of OPA 90—the long-term prognosis may still be bleak, unless ways can be found to train salvors and stimulate salvage-related business activity.

3

NATIONAL SALVAGE POSTURE ISSUES

The regional assessments in Chapter 2 point to a number of broad issues associated with salvage readiness and capability. These concerns are fundamental to the development of a comprehensive understanding of marine salvage in the United States. This chapter analyzes these issues. They are:

- Salvage business conditions
- Decision making in salvage incidents
- National salvage policy
- Salvage human resources
- Rescue towing
- Marine firefighting
- Cargo transfer, including lightering
- Damage stability information
- Salvage in the presence of hazardous cargo
- Jettisoning
- Safe havens
- Salvage readiness of vessel and crew
- The contribution of the U.S. Navy in national salvage response

SALVAGE BUSINESS CONDITIONS

Salvage as a commercial effort traditionally has focused on the saving of property—ships and cargo. The salvor's ability to make money depends on the commercial climate, the competition, the salvage contract, and government regulation. Salvage capability in an area depends on demand, which in turn hinges on vessel casualties. The International Salvage Working Group reports that over the last 12 years, vessel casualties have declined annually worldwide, despite increasing numbers of ships at sea (Tecnitas, 1992). This trend is due to significant improvements in marine safety, and if it continues, it can only make it more difficult to sustain commercial salvage capability. Demand for services will not produce enough income for salvors to operate without other, more dependable sources of income, such as ocean towing or marine engineering. Such activities have become salvors' core businesses, while salvage is undertaken only as opportunities arise. The problem with this practice is twofold: it diminishes the salvage award, and it tends to discourage salvors from remaining in the business. These points are explored later in this section.

The salvage industry encompasses three more or less distinct segments:

1. *Full-time professional salvors* maintain organizations, personnel, ships, and specialized equipment for salvage. Professional salvors range from large companies with considerable capital investment and widespread operations to smaller businesses with limited resources serving restricted geographical areas. The latter are most common in the United States. Ships equipped for

salvage are extremely important, but the dedicated salvage ship waiting on station is no longer the hallmark of the professional salvor. Instead, professional salvage companies have experienced workers available on call and a pool of specialized salvage equipment and are willing to drop routine work to handle emergencies.

2. *Independent salvors* are often one-person operations and vary widely in knowledge and experience. They seek opportunities for salvage but maintain little equipment, preferring to rent it for each job. Often they advise the shipowner on equipment that should be obtained for a specific casualty.
3. *General marine contractors* may provide salvage on an ad hoc basis, pursuing salvage work in a local area but not dependent on it. Such contractors are used mainly for non-time-critical salvage, such as wreck removal and marine repair. They may supplement their resources by subcontracting with other salvage providers in the region.

The use of independent salvors and general marine contractors, in combination with the declining casualty rate, have put immense financial pressure on the professional salvor. Independent salvors and general contractors often charge shipowners less than professional salvors because they don't have high overhead costs for specialized salvage equipment and personnel. Moreover, independent salvors and general contractors take the lower priority harbor clearance work that used to be the professional salvor's bread and butter, offsetting the expense of maintaining otherwise underutilized equipment.

Revenues generated by such ships often do not offset costs, which have driven salvage ships from their former stations at key points in shipping lanes with a history of accidents. As an alternative, salvors have developed rapid response capability, drawing on experienced, trained personnel and equipment from central bases and strategically located equipment stockpiles. Salvors often find it practical, expeditious, and economical to charter needed ships and lease common equipment locally, so that they purchase and carry only specialized equipment. By changing their mode of operation to meet altered conditions, professional full-time salvors have narrowed the gap between the two. This shift also has blurred the distinction between the professional and the independent salvor. In both the *Mega Borg* and the *Exxon Valdez* incidents,[1] the salvors (or salvage company) had years of experience but were viewed as operating as independent salvors.

The growing emphasis on environmental protection complicates the salvage business, but doesn't necessarily produce additional income (NRC, 1994). The traditional criterion for success in salvage has been the saving of property. The definition of success has changed as environmental protection has assumed increasing importance. Indeed, environmental concerns often govern operational decisions during salvage, and environmental protection has become—in addition to the saving of property—a determining factor in the success of salvage operations. Frequently, environmental protection is best achieved by saving the property in peril—the vessel.

Commercial salvage transactions must be covered by a contract that is fair to all parties and ensures adequate rewards for the service provider. The "no cure, no pay" open form contract has been most common in the international salvage industry. A number of these contracts exist: Lloyd's Standard Salvage Agreement, frequently called Lloyd's Open Form (LOF), is the most common and best known. This type of contract is not favored by American salvors, particularly those on the West Coast, who may deploy equipment long distances and, on arrival, find the casualty so deteriorated

[1] All instances of salvage in U.S. waters reported to the committee are listed in Appendix E.

that no "cure" is possible.

In addition to encouraging the maintenance of professional salvage organizations, an open form contract provides a set of terms without addressing specific compensation. The terms are well known, reducing any need to haggle over details and delay the work. Shipowners often feel the open form contract favors salvors, but this is not the case. Such a contract is a high-risk agreement for salvors, especially with regulators playing an expanding role in salvage operations.

Under the LOF and other open form contracts, the salvor's award is determind by arbitration. Although arbitrators may be experienced and knowledgeable regarding salvage, the amount of an award is not predictable. The U.S. Supreme Court, in the case of *The Blackwall,*[2] set forth criteria for determining the proper amount of a salvage award. In brief, these are:

1. Labor expended by salvors
2. Promptitude, skill, and energy in rendering services
3. Value of property employed
4. Risk incurred
5. Value of property saved
6. Degree of danger from which the property was rescued

The *Blackwall* decision may have binding value as a precedent in U.S. courts only; nevertheless, its criteria are generally followed by arbitrators in establishing awards both in foreign and U.S. courts.

The courts have held that a salvor who maintains expensive equipment and personnel solely to render salvage services is entitled to a greater award than other responders. But in the 1990s there are so few calls for such services that it seems counterproductive to penalize professional salvors who perform work other than salvage. This reality has been recognized to some degree. The U.S. Court of Appeals for the Second Circuit has stated, for example, that "exclusive devotion of a company's resources to salvage is not a condition precedent to recognition of a professional salvor's favored status,"[3] and courts and arbitrators have just begun to recognize "part-time professional salvors" as "deserving of an increment due to their availability to persons in distress."[4] But the salvage industry—which cannot rely solely on salvage and must accept other work—is nearly unanimous in its view that unless a contract specifies an amount, there is no assurance the award will be adequate.

Accordingly, no-cure-no-pay salvage contracts are falling out of favor, not only because the awards may be inadequate but also because arbitration is often lengthy. While salvors and other claimants may feel that "justice delayed is justice denied," some progress has been made in addressing this complaint. LOF-90 has introduced steps to cut delays in arbitrations, and arbitral procedures under the Japanese Shipping Exchange Agreement are deliberately short, so that the salvor receives a prompt award.

When compensation depends on successful results, any work carries high risk. Salvors can spend large sums and receive nothing in return. Yet salvors faced with an impossible or unprofitable job cannot simply declare "no cure" and walk away. They must follow established procedures to be relieved of their contractual obligations, while the salvage activities, costs, and losses continue.

Public concern for environmental protection and a parallel increase in environmental legislation exacerbate the uncertainty of the business aspects of salvage. Regulatory intervention that may occur outside the business framework of

[2] 77 U.S. 1, 10 Wall.

[3] *B.V. Bureau Wijsmuller* v. *U.S.*, 1980 AMC 2331 at 2356.

[4] *Bindon* v. *Jones*, 1986 AMC 1403 at 1405.

salvage adds a factor the salvor must consider in protecting income. A salvor taking an open form contract in today's liability environment takes an undefined risk—one salvors rarely can afford. The risk of pollution is so great that the no-cure-no-pay provision of the LOF was changed in 1980 to provide a safety net for salvors. This revision allows for departure from the strict no-cure-no-pay principle in some cases by assuring that salvors be compensated for costs incurred to minimize pollution from a casualty. The International Salvage Convention of 1989 broadened the importance of a salvor's efforts to minimize or prevent pollution by including those efforts as a criterion for setting awards. In addition, the convention provided that, if the vessel was lost but efforts to prevent or minimize pollution were demonstrable, the salvor could be eligible for expenses, an additional award up to 30 percent of expenses, and—in special circumstances—100 percent of expenses. In addition, a forthcoming International Salvage Union standard wreck removal contract changes the terms from fixed price, no cure no pay, to fixed price with payment for progress. Even with these changes, a perception exists that the traditional open form contracts, with award based on salved value, are inappropriate given that the measure of success is changing to emphasize environmental protection.

For all the above reasons, the no-cure-no-pay principle is losing viability. Most American salvors prefer to work under a daily rate contract, often with an incentive bonus, such as a percentage of salved value. These contracts protect salvors' income and eliminate many problems associated with open form contracts and arbitrated awards. They are safe even under OPA 90, because the salvor is paid for services provided regardless of success or failure.

When conditions combine to reduce profits to the point where they are no longer satisfactory, salvors leave the industry. This has occurred to some extent in the United States. The system of retainers provided by the tanker industry to ensure the availability of salvage resources as required by OPA 90 has created a new revenue source for salvors—a retainer to ensure the provision of salvage resources in the future. For the majority of individuals and companies viewing themselves as professional salvors, such a retainer may be the factor that convinces them to remain in the business. The long-term effectiveness of such a system in maintaining professional salvors has yet to be proven, since retainers are neither required nor standard practice. The oil-transport industry will continue its search for the lowest-cost option in fulfilling its responsibilities under OPA 90 and other regulatory mandates. However, if more marine transportation companies act as their own salvors, providing retainers only for companies with special expertise, the influence and viability of the traditional salvors may be threatened further.

Findings

The salvage industry is undergoing major changes, which are driven by three primary factors:

1. The industry has shifted from servicing the private sector, by saving vessels and cargo, to a public service role of preventing casualties from damaging the marine environment and the local or broader economy.
2. The business structure of the industry has changed from full-time dedicated salvage companies to organizations for which salvage is either a segment of their overall business or coincident to other roles in general marine contracting.
3. Vessel casualties have declined, and there has been a corresponding reduction in opportunities to earn salvage revenues.

These factors continue to depress the salvage industry, along with traditional compensation practices that have not kept pace with the risks and costs associated with salvage, and which are only recently undergoing change.

The effect of OPA 90 on the salvage industry is uncertain. The regime of retainers fostered by regulations arising from that law will provide additional income for at least some segment of the salvage industry. But whether that revenue stream is sufficient to further improve the economic viability of the salvor and the marine salvage posture of the United States is not yet known.

The improvement in salvage capability due to OPA 90 has been supported solely by the tanker industry. This approach places an undue burden on one segment of the marine industry to provide resources that may be tapped by the entire industry and may not produce sufficient revenue to maintain current capabilities over the long run.

The current liability atmosphere, charged by the threat and expense of marine pollution related to a vessel casualty, has significantly increased the salvor's financial risks and the costs of providing salvage assistance. The standards for compensation have not kept pace with these increased risks and costs.

SALVAGE DECISION MAKING

A marine casualty that requires salvage services is an emergency. The need for clear, positive decision making continues throughout the emergency, and there is no room for error, equivocation, or delay. The consequences of faulty decision making can be catastrophic. These consequences alone demand a tough approach to decision making.

Three influences work against logical and methodical decision making:

1. The situation is dynamic. Many of the dynamic influences—notably weather—are completely beyond the control of personnel on the scene.
2. A clear and complete picture of the situation is not available in the beginning. The picture develops as the operation proceeds, and in some cases it is never complete. Accordingly, salvage plans and decisions are made with limited information and are modified and updated as new information becomes available.
3. Casualties usually deteriorate over time. It is not possible to delay decisions without increasing the risk to the vessel and the risk of side effects, such as pollution.

Analysis of and response to a casualty require expert knowledge applied expertly. Salvage decisions are time-critical, on-the-feet judgments, requiring practical knowledge of seamanship, seat-of-the-pants engineering, and the ability to analyze, extrapolate, and decide. The expertise needed to make salvage decisions comes only from experience with casualties in the field.

Knowledge and experience with vessel casualties is not widespread in the marine industry—most mariners go through their entire careers without experiencing a major casualty. Coast Guard officers and other regulators may acquire broad knowledge of maritime matters, but may not have much experience with marine casualties due to changing assignments and extensive collateral duties. Representatives of environmental agencies are generally in very unfamiliar territory in dealing with a casualty. Salvors, surveyors, and others who deal with casualties regularly are usually best equipped to make sound salvage decisions.

Traditionally, salvage decision making involved only the vessel owner, the underwriters who had an interest in the cargo and the vessel, and the salvor. The owner and underwriter usually deferred to the salvor, once hired. Other parties were not represented, even though the interests of the general public and the environment were affected by the decisions made. That small decision-making team, able to act quickly and authoritatively, has been replaced in incidents where environmental or public harm are a potential outcome by a partnership of the Coast Guard, the affected state(s), and the shipowner's predesignated representative. As a result, the decision-making process now incorporates the interests of a number of additional affected parties, and may not be able to respond as quickly and directly as in previous times.

Figure 3-1 shows the partnership regime that exists under the National Response System (NRS) and the Unified Command System (UCS). The NRS is a hierarchy of planning and response teams composed of representatives from federal, state, and local agencies (Johnson, R., 1993). As the figure shows, these agencies have input into the NRS through a variety of mechanisms, but actual field operations are directed by the federal on-scene coordinator (FOSC), the owner's designee (referred to in the Figure as RPOC, or responsible party on-scene coordinator), and representatives of the affected state(s). This on-site decision-making process constitutes the UCS. While this partnership was designed for incidents that threaten or actually produce pollution, the response to any major vessel casualty may be organized in the same manner. This organization, modified from the Incident Command System widely used in firefighting and adapted to marine casualties, is in direct contrast to the former decision-making process in which the salvage situation was negotiated solely between the salvor and the vessel owner, with the overriding concern the saving of the vessel and the cargo.

In recent years, regulatory authorities, particularly the Coast Guard, appear to have gained so much legislated power that the vessel owner and salvor may feel the partnership is nonexistent or at least lopsided. The relationships among the stakeholders—ship and cargo owner, insurers, the general public, and the environment—have not changed; only the manner of decision making has changed.

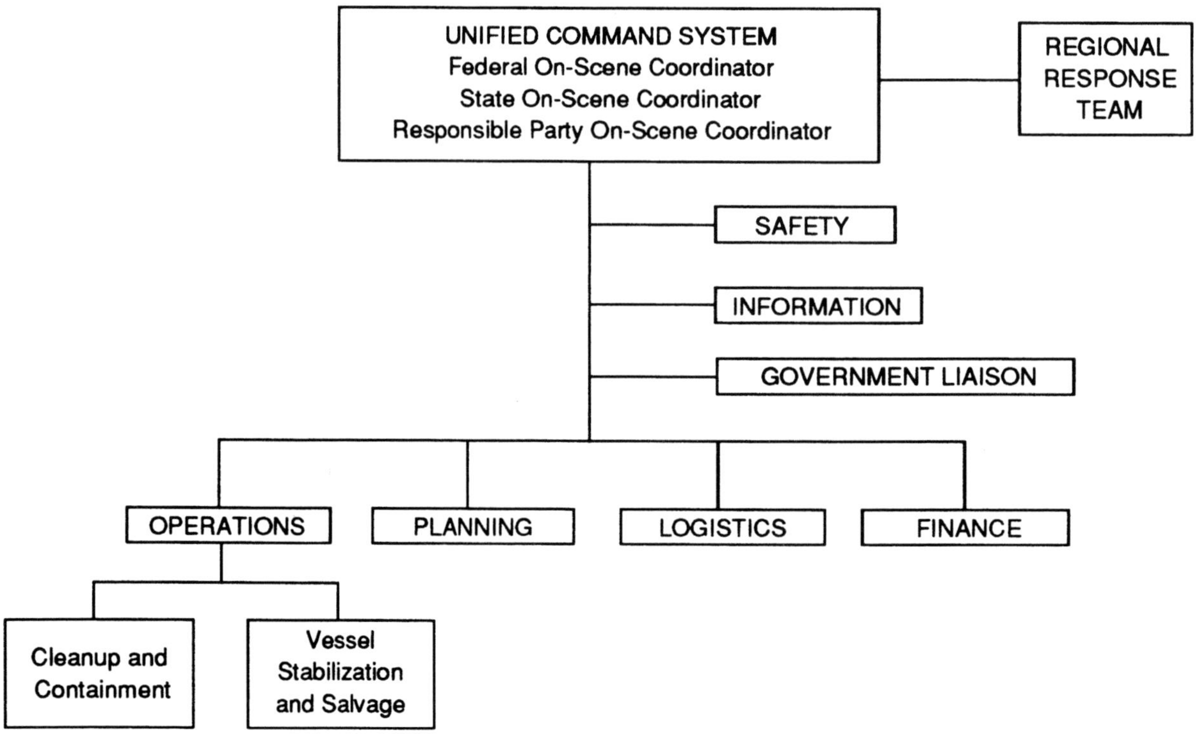

FIGURE 3-1 Unified command structure.

However, adding stakeholders to the decision process has the potential to have a significant effect. At best, time passes during the discussions and debate, and efforts may be diverted from solving the problem. At worst, inappropriate directed action may eliminate viable options and degrade overall operational effectiveness. On the positive side, however, the vetting of "what if" alternatives may lead to a solution that, from the perspectives of all stakeholders, provides superior overall response, even if delay allows the problem to worsen in the interim.

The involvement of the Coast Guard and others appears to be the rule rather than the exception in incidents involving actual or potential pollution. In the recent cases of the *Mega Borg* in the Gulf of Mexico and the *Ocean 255* in Tampa Bay, the Coast Guard asked the Navy to provide standby support. In both instances, salvors under contract to the Navy provided oversight and consulting services to the Coast Guard during the salvage effort. In the case of the *Ocean 255*, the Coast Guard strike team performed the actual cargo removal, with the support of outside contractors. The Coast Guard is directed to "shoot first, and ask questions later" in responding to casualties involving potential or actual pollution, directing and prescribing specific action. The record to date of Coast Guard decision makers in salvage situations occasionally has led to discussions about the agency's unilateral decisions regarding salvage response without due regard for contributions of the vessel owners and their salvors. However, until the policy results in intolerable interference, no change is expected.

In addition to the expanded decision-making team, other factors complicate emergency response. In particular, the proliferation of environmental regulations and the variations among them in different jurisdictions complicate salvage decision making. An operationally sound decision may violate one or more environmental regulations and could lead to criminal or civil penalties. Moreover, even the *threat* of environmental penalties may force the choice of operationally and technically poor options. Attorneys properly act as advocates for their clients and represent their interests, but the advocacy of a particular position during a dynamic emergency can delay the process and divert decision makers from the soundest course of action. Media attention, perhaps founded on less-than-expert technical understanding, may shift public perception away from the true situation and heighten public pressure for a specific action.

Sound salvage decisions seldom affect all parties alike. Decision making has changed because of the need to ensure that all parties' interests are addressed and conflicts minimized. The Coast Guard has met this need, at least in potential pollution incidents, by incorporating the UCS. This system is based on the premise that all parties must contribute to the strategy, planning, operations, and use of required resources, and, therefore, that a decision-making partnership is needed involving the FOSC, the responsible party, and the affected federal and state jurisdictions. The UCS is intended to allow for the FOSC to resolve disputes, ensuring that decisions are made in a timely fashion. The UCS is a developing concept, which will change and evolve with experience. The mandate for the marine industry is to be involved in that process, and, to the extent that salvage is part of response planning, to get the salvage industry involved as well.

Findings

Decision making in response to vessel casualties, once the purview of salvors, owners, and insurers, has changed significantly, particularly when it relates to casualties with the potential for marine pollution. Decision making at the site of a casualty now relies on the consensus of a unified command, consisting of the federal government, the responsible party (the owner or designee), and the affected state government(s).

Transferring authority to a committee may delay time-critical salvage decisions and tends to restrict the salvor's actions. Salvage plans now need to be vetted through the UCS, and any deviations from agreed-upon plans of action are performed at the peril of the executing party.

NATIONAL SALVAGE POLICY

The 1982 NRC report concluded that "The Congress should update the national statement of salvage policy (10 USC 7361–7367, Salvage Facilities Act of 1948) to recognize the vital role that salvage plays in minimizing the public consequences of maritime casualties, and to harmonize the Salvage Act with the laws on intervention and pollution response" (NRC, 1982). Despite that recommendation, there has been no change in the national salvage policy in the last decade.

Current Statutory Status of Salvage

The current national salvage policy, as contained in the Salvage Facilities Act, may be summarized as follows: "The public interest is served by maintaining salvage capability to provide for the national defense, especially to ensure readiness for war mobilization" (NRC, 1982). A number of other U.S. statutes also pertain to salvage. These include "Cabotage Law" (Act of June 11, 1940), the Saving Life and Property Act of 1949, Waterways Management Act, the Clean Water Act of 1972, the Intervention on the High Seas Act of 1974, and most recently, the Oil Pollution Act of 1990 (OPA 90). (These laws are summarized in Appendix H). Each of these statutes contains specific references to various aspects of salvage, but none clearly outlines a concise statement of national policy as it relates to salvage. Moreover, none of the statutes states explicitly that an additional function of salvage is to minimize pollution (although that responsibility is implied), and none addresses the public safety role.

The most concise, general statutory statement of U.S. salvage policy is found in the Salvage Facilities Act. This act was promulgated following World War II, "to authorize the Secretary of the Navy to provide salvage facilities and for other purposes." While there has been no court decision interpreting this act, the legislative history shows that Congress had several objectives:

- To provide salvage resources to protect the redeployment of government-owned war materiel on chartered ships (but not in excess of national defense needs).
- To foster (but not necessarily subsidize) the commercial salvage industry.
- To allow (but not require) the Navy to render salvage services to private vessels when commercial salvors are not available, charging for those services to support the Navy's salvage facilities.

This statement of policy has not been addressed or reissued since its promulgation in 1948 and does not reflect either the current international situation as it pertains to national security or the current commercial salvage situation.

Environmental aspects of salvage policy are suggested in OPA 90, which addresses limited salvage functions carried out to prevent or minimize pollution. These provisions, limited to vessels carrying oil as cargo, have led to the adoption of Coast Guard regulations requiring such vessels to ensure the presence of lightering and firefighting capability and salvage support by "contract or other approved means"[5]

[5] 33 CFR Part 155.1035 (d)(6).

as part of their federally required contingency and response plans. Some states, specifically Washington, Virginia, Florida, and Texas, apply the requirements for contingency and response plans to a broader class of vessels, including those carrying significant amounts of fuel.

The United States has ratified the International Convention of 1989, which recognizes that one of the goals of successful salvage is to prevent or minimize damage to the environment from vessels or their cargo by incorporating the salvor's skill and efforts to do so into criteria for determining salvors' awards. The convention does not require that the vessel or cargo be saved for the salvor to be eligible for compensation for efforts to prevent or minimize environmental damage. Although the convention is not yet in force internationally, its ratification domestically defines the direction Congress expects national policy to follow.

Current Government Salvage Policies

The role of government agencies in salvage has not been defined explicitly, although a number of agencies, notably the U.S. Navy, Coast Guard, and Army Corps of Engineers, play an active role in salvage response. In accordance with the Salvage Facilities Act, the Secretary of the Navy has entered into competitively awarded contracts with three commercial salvors for services beyond the Navy's capabilities. These contractors are compensated for their costs for operations conducted. Additional salvage-related contracts are awarded to companies that provide limited services using equipment owned and furnished by the Navy. In summary, through the Navy, the United States provides limited support to the commercial salvage industry through standing salvage contracts, contracts for salvage-related services, and leases of salvage and pollution abatement equipment.

The Coast Guard, in addition to providing an FOSC for major vessel casualties, maintains three regional strike forces, on the Atlantic, Gulf, and Pacific coasts. These assets are available at the request of a FOSC to assist in marine casualties where pollution mitigation is a concern. The FOSC also is authorized to request the assets and contracts of the Navy's supervisor of salvage, to further assist in mitigating impacts of marine casualties.

The Corps of Engineers is responsible for maintaining the waterways of U.S. ports and can provide assets for spill response. For example, floating cranes used to remove wrecks that pose hazards to navigation or block channels and dredges can be used in oil spill response. A dredge was pressed into service in the *Exxon Valdez* case.

It is the policy of government agencies to utilize government assets to help private vessels only when commercial assets are not available. However, in incidents that threaten or cause marine pollution, the liability implications of OPA 90 and Congress' expressed desire for rapid and effective response have induced the Coast Guard to take a proactive approach—to respond immediately and sort out the consequences later. In recent cases, such as the *Mega Borg* and the *Ocean 255,* the Coast Guard mobilized its own equipment and requested Navy support within hours of the accident. Navy salvage contractors requested by the Coast Guard were on the scene within six hours in the latter incident and, as noted earlier, the Coast Guard strike team performed the cargo transfer and subsequent dewatering of the cargo tanks in concert with a salvage company.

National Salvage Policy Considerations

Since the 1982 NRC report was published, several trends have emerged that serve to reinforce its conclusions. Among these trends are:

1. The aforementioned shifting emphasis in salvage from private service to a more public or societal role of protecting the environment and the economy.
2. Heightened public awareness of the adverse impacts of pollution from vessel casualties such as the *Exxon Valdez*, and the translation of such awareness into broad congressional initiatives such as OPA 90.
3. Recognition of potential economic losses resulting from disruptions of port activities due to a marine casualty.
4. Changing military needs stemming from the end of the Cold War.

The need to ensure both a consistent federal approach to salvage and timely and effective response is becoming more apparent. In the *Exxon Valdez* case, the need for combined private and federal resources, including the Coast Guard strike team, Navy pollution response equipment, and various Department of Defense (DOD) aircraft for transportation, demonstrated that interaction between the public and private sectors is necessary to respond to a major pollution casualty. Likewise, the Persian Gulf war confirmed the need to maintain, through the Secretary of the Navy, a significant salvage posture capable of functioning throughout the world to protect national security. Finally, recent incidents involving the closure of the Houston Ship Channel for several days demonstrated the potential for adverse economic impacts that is neither effectively nor rapidly addressed by either the public or the private sector consistent with an integrated national policy.

It is clear from the definition of salvage used in this report (see Chapter 1) and the 1982 NRC report that all elements of the marine commerce system—whether public or private assets, oil or hazardous material carriers, or bulk and cargo carriers—may be affected by marine casualties and therefore may benefit from having a comprehensive salvage capability available in U.S. waters. As a result of recent legislation, primarily OPA 90, changes in salvage capability are being financed in part by oil carriers through a system of retainer agreements. Such regulation of one segment of the marine commerce system, while it may improve salvage capability, focuses only on pollution response and therefore may not provide for the type of comprehensive salvage policy the nation requires.

National salvage policy requirements have changed from those stated in the Salvage Facilities Act to include the following goals: protect the national security; minimize or prevent environmental impacts due to pollution from marine casualties; protect public safety; and ensure minimal disruption to the U.S. economy resulting from marine casualties in the nation's ports and waterways. A number of alternatives were considered for updating the national salvage policy. For example:

- Develop a national, publicly funded marine salvage capability patterned after those in countries such as France and South Africa. (A similar capability supported by the U.S. Navy existed in the United States during the 1950s and 1960s but was terminated in the late 1970s due mainly to economic viability.) The fiscal implications of such a commitment may rule out this alternative, given the current political and economic climate.
- Rely on the regulatory mandate that compels one segment of the marine community—the liquid cargo carrier—to provide private support through contract or other means for selective salvage capabilities. This alternative may place a disproportionate burden on one segment of the industry to provide salvage capability for the overall industry and the public good.
- Expand the regulatory mandate to encompass all significant elements of marine commerce. This change would require specific statutory authorization, but it may be consistent with the principle of fairness by suggesting that all the elements posing the risk should shoulder the burden of protection from that risk.

- Place stronger restrictions on the use of government resources in salvage incidents, with the expectation that the private sector would have more incentive, both in terms of financial rewards and response needs, to be prepared to respond to marine casualties. This alternative may be inconsistent with both the direction of the Congress in OPA 90 and the economic conditions in the marine salvage industry.
- Codify or clarify the essential salvage roles of government in minimizing public consequences to the environment, the economy, and public health and safety in conjunction with meeting national security needs. At the very least, the policy would require a clear definition of government agency roles and their relationships to the private sector. The policy should clarify the following two functions. First, in marine casualties where life, the environment, or other public interest is threatened, the U.S. Coast Guard is responsible for oversight and/or direction of the response to the casualty. Second, the U.S. Navy should develop technical programs advancing salvage response capability and equipment, provide technical assistance in nongovernment marine casualties requiring salvage response when requested by either commercial and/or government agencies, contract for salvage services in excess of internal capability (and make such services available on request), and provide training for U.S. Navy and other personnel. A clear statement of noncompetition with private resources would also be required.

Implementation of any of these alternatives would have to address specific elements of salvage operations, such as rescue towing, firefighting, lightering, adequacy of training, minimum qualifications of salvage personnel, wreck removal, deepwater search and recovery, and harbor and channel clearing.

Findings

There is no clear, concise statement of national salvage policy that addresses the role of salvage in minimizing the public consequences of marine casualties. Various federal laws, statutes, and regulations imply that salvage plays a part in protecting the public, but that role is not stated explicitly. Absent a stated policy, the Coast Guard and the Navy have assumed significant roles in management and oversight of marine casualties, particularly those involving pollution and the threat of pollution, through interpretation of their roles arising from Congressional mandates such as OPA 90.

The primary motivation for maintaining a salvage capability has shifted from concern for private assets—the vessel and cargo—to a more public or societal interest in protecting the environment and the economy from the impacts of a vessel casualty.

Current improvements and changes in salvage capability are largely driven by regulations requiring vessels carrying oil to ensure that selected salvage capability is available in the event of a marine casualty that involves actual or threatened pollution. These changes are directed at mitigating the risk of pollution and do not constitute the broad-based improvement that may be needed to address all public concerns. The costs of implementing the changes are born by only one segment of the marine transportation industry.

THE CONCEPT OF THE "PROFESSIONAL SALVOR"

The 1982 NRC salvage report observed that "[m]ost of the few companies that conduct marine salvage in the United States see salvage operations as supplements to their primary operations of marine transportation, point-to-point towing, or marine engineering. Time-critical salvage incidents occur too infrequently and sporadically, and remuneration is too low, to justify single-purpose salvage companies or maintaining specialized vessels for salvage as in the past . . . " (NRC, 1982). That situation has not changed.

The composition of the salvage industry has shifted over the past several decades. The industry once was composed of, even dominated by, a relatively small number of traditional or dedicated salvors, whose only—or at least primary—business was ocean salvage. These companies were integrated, self-contained operations that owned large offshore salvage vessels and large stocks of specialized salvage equipment and employed dedicated salvage personnel. In the last decade, two of the companies that were active in 1982 have ceased to conduct salvage operations, and the transformation of the salvage industry, accelerated by OPA 90, is being shaped aggressively by new entrants.

The term "traditional salvor" has lost its meaning in the United States. Indeed, it is difficult for many in the marine industry today to even define salvor. The discussion of salvage business conditions earlier in this chapter pointed to a blurring of the distinction between a professional salvor and an experienced marine contractor conducting time-critical salvage.

The situation is similar to the one that existed in the oil spill response industry prior to the enactment of OPA 90. At that time, oil spill response contractors needed little more than a few feet of boom to be listed in contingency plans that were intended to serve as guides in the event of a spill. Regulations issued pursuant to OPA 90 require that a company seeking to be qualified as a spill response contractor meet criteria that serve as a minimum standard for responders.

OPA 90 does not require the Coast Guard to promulgate regulations stipulating the criteria to be met by salvage companies as it does for oil spill contractors, but nothing in the law precludes the Coast Guard from doing so. As yet, the Coast Guard has not issued criteria for qualifications for the salvage industry. As a result, no such criteria exist for determining whether a professional marine contractor or service company has the experience and wherewithal to mount a major salvage response. Salvage is a complex operation, consisting of many diverse elements, and the procedures change depending on the environment. As defined in Chapter 1, salvage includes:

- Wreck removal
- Harbor clearance
- Deep-sea search and recovery
- Firefighting
- Refloating a vessel
- Offloading cargo
- Shoring, patching, and making temporary repairs
- Rescue towing

All or only one or two of these activities may be required in any salvage response, and, depending on the situation, they may be performed by different companies coordinated in a single effort. The development of criteria for defining "salvor" would require acknowledgement that not all contractors can or want to perform all the elements of salvage in every region of the United States, and that any contractor may be extremely capable, and even preferred, for certain aspects of salvage or in certain regions. The expertise and resources needed to respond to a casualty on the western rivers, for example, are undoubtedly different from those required in an offshore casualty scenario.

The Navy, in its contracting for salvage support, establishes criteria for qualifying bidders, addressing both company type and the nature of personnel experience. The Coast Guard does not have similar criteria to judge contractors who may be employed in salvage response it manages, or to advise shipowners concerning which contractors would be acceptable.

A first step in determining what constitutes a professional salvor would be to develop a list of qualifications, similar to those used for oil spill responders, for determining active salvors. The list should include salvage companies that have the capability to handle an entire major salvage operation and other companies that specialize in a specific marine salvage element(s). The companies on the list should have—at least—the following attributes (no attempt has been made to quantify them):

- Current activity in one or more aspects of the marine salvage industry.
- A history of participation in successful marine salvage operations.
- Significant investment in salvage equipment.
- Qualified personnel experienced in marine salvage on staff or otherwise available for deployment within a specified response time.
- Ongoing training programs to improve salvage skills among personnel.
- Realistic salvage plans for the types of marine salvage in which certification is claimed and for the geographic areas of interest.
- Ability to undertake time-critical response to marine emergencies.

These general criteria need to be further developed, and specific criteria need to be determined for each of the various aspects of salvage listed earlier, along with modifications for various regions and marine environments. These criteria must be developed by independent bodies, rather than the industry itself, to ensure public and commercial acceptance. Otherwise, the criteria may be viewed as self-serving for only a few companies and may not be accepted as a viable standard. Obvious candidates for undertaking the task of developing cogent and acceptable standards are the Navy, given its experience and involvement in salvage, and the Coast Guard, which is often responsible for overseeing response to incidents in which marine salvage capability is exercised.

The criteria could be used to identify active salvors for inclusion in contingency plans or to respond to public or private requests for recommendations concerning salvage companies. The use of professional salvors identified by the established criteria would be an obvious means of improving the business climate for professional salvors and eventually enhancing the investment in and improvement of salvage readiness.

Findings

The changing composition of the salvage industry has continued, with economic concerns dictating that the professional salvor rely increasingly on outside support contractors due to the high overhead of maintaining a completely integrated operation. The general marine contractor continues to play a major part in salvage, particularly in non-time-critical salvage operations, such as wreck removal, which used to be the mainstay for professional salvors.

Specialized firms focusing on one aspect of salvage, such as firefighting or lightering, and featuring fly-away or prepositioned salvage systems, are becoming increasingly common. These firms have a national focus, providing specialized equipment and trained personnel to casualties anywhere in the United States.

No standards exist that can be used by industry, regulators, or the public to determine the capabilities of companies actively offering salvage services. In some cases, this lack of criteria imposes an additional burden of oversight on individuals or organizations assigned to monitor a salvage operation; inexperienced salvage providers, relying on a low bid to obtain work, may not perform to the needs of the situation.

SALVAGE HUMAN RESOURCES

Trained, experienced, and motivated personnel are essential to successful salvage. Like many endeavors, salvage is best learned through experience. Education, experience, and salvage training are helpful, but real salvors are made on the job. The more jobs, the more experience; the more experience, the greater the likelihood of success.

While salvors generally are trained on the job, the U.S. Navy has formal training programs to meet its own needs. But due to the dispersion of personnel and the paucity of training opportunities, only a small number of the Navy trainees become seasoned salvors, and even fewer find their way into the commercial salvage industry. Although the availability of salvage personnel did not appear to be a problem at the time of the 1982 NRC report, the lack of on-the-job training opportunities and the declining numbers of Navy salvage trainees were noted.

Because the salvage industry has developed few new experienced personnel since 1982, the people available (even on call) undoubtedly have declined, and their age is advancing rapidly, a concern echoed worldwide (Tecnitas, 1992). Furthermore, the output of Navy training programs is barely 50 percent of what it was prior to the breakup of the Soviet Union.

The disappearance of dedicated salvage ships has been counterbalanced in part by the ingenuity and good work of independent salvors, marine contractors, and towing firms. But there is little opportunity for salvage team training, on the job or otherwise, and no motivation for such training within companies that are not committed to maintaining a salvage capability. Even when a company keeps a specialized salvage ship, the committee found that personnel assigned to the ship have fallen to as few as two crew members. While these individuals may well constitute the nucleus of a salvage team, a complete team likely will be formed piecemeal, using anyone who can be spared from full-time assignments in nonsalvage work. There is a contention within the industry and among observers that the quality of salvage efforts will suffer, perhaps imperceptibly now, but increasingly so as the current generation of experienced salvors reaches retirement (Smith and Reed, 1990).

There is no easy answer to this problem. The training ground for salvage is the casualty, and as the casualty rate has declined the opportunities for training and maintaining skills have dwindled. In addition, with the vessel owner sometimes managing its own salvage response, occasions for professional salvors to expose their personnel to even the infrequent casualty are limited.

If the traditional salvage industry continues to deteriorate, then the availability of competent personnel will decline also, unless the marine transportation and salvage industries commit to a focused program to improve career prospects for new entrants and to increase formal and hands-on training for existing personnel. This issue also requires attention from government, because it may not be reasonable to concentrate new training programs within private salvage organizations in today's high-technology environment. These organizations may not have all the requisite expertise in-house; even the professional salvor, once almost self-contained, relies more and more on outside specialists for salvage engineering, firefighting, lightering, naval architecture, and the provision of the salvage working platform itself.

The present lack of training opportunities for salvage personnel—masters, engineers, divers, and mechanics—is not expected to improve. Nonetheless, there are steps that could be taken to arrest the declining numbers of trained salvage personnel. The first would be to develop an industry-supported curriculum at one or more of the maritime or training schools, much like the maritime fire-training schools. The curriculum would have to provide theory and hands-on experience in various phases of salvage, which would require significant funding for training "casualties." Maritime schools are beginning to consider salvage-related curricula (Ringelberg and Banks, 1993).

A further approach would be to open the Navy training facility in Panama City to selected industry and government (state and federal) personnel. This measure would have a number of benefits:

- It would make use of the training command, which has incurred cutbacks as a result of the reduced military threat in the post-Cold War world.
- It would foster public/private cooperation in case of a national emergency or a major commercial casualty through the common training of industry and Navy personnel.
- It would provide industry input into the training program, thereby enhancing the preparation of Navy personnel for civilian careers and providing a source of future personnel with skills appropriate for the salvage industry.
- It would allow common training for federal and state personnel who may be tasked with oversight of commercial casualties requiring salvage activities.

Training costs could be reimbursed by trainees, their companies, or organizations, or financed under the Salvage Facilities Act. Curriculum modifications would be required to focus on the commercial aspects of salvage.

Findings

The base of experienced salvage personnel continues to decline and age with the reduction in numbers of marine casualties and the dispersion of salvage opportunities among various segments of the marine contracting industry. The salvage industry is attracting few new individuals because there are limited opportunities to participate in actual casualties and few chances to advance in a declining industry.

On-the-job training is extremely limited due to the declining casualty rate, and salvage expertise cannot be acquired without hands-on experience. The loss of experienced personnel can be expected to continue unless there is active intervention by the salvage and marine transportation industries, and perhaps the government, to provide training and career path enhancement.

RESCUE TOWING

Rescue towing may best be defined as the ability of a rescue vessel initially to keep an incapacitated vessel out of harm's way and ultimately to deliver it to a safe haven. A rescue vessel may be a dedicated salvage vessel, an offshore supply boat, a tugboat, a commercial vessel, or a government vessel. It would be misleading to define a rescue vessel as having a specific feature or task. Economics drives the direction and vigor of salvage, as it does other businesses. Because there are far fewer casualties today than there were even 25 years ago, dedicated salvage vessels simply cannot amortize the high costs of construction, maintenance, and full-time skilled salvage crews. The result is that there are very few dedicated salvage tugs on station in the world today that operate without government subsidy (Milwee, 1993; Tecnitas, 1992).

In the United States, rescue towing has been carried out by high-horsepower tugs normally engaged in the commercial practice of moving vessels in coastal trade or supporting the offshore oil and gas industry. These tugs may be diverted, depending on their availability, in emergency circumstances. Many towing and offshore supply companies contract these tugs by the hour or day. A rescue contract may be negotiated by the shipowner and tower at close-to-normal towing rates if, for example, a ship experiences steering failure and/or machinery breakdown away from shore under fair weather conditions. It appears that the system has worked fairly effectively in the past. There does not appear to be a recent case in the United States in which a casualty has failed to receive the assistance requested.

As the distance from shore increases, response capability diminishes, due to the geographic distance and to the occasional reluctance or even refusal of coastal tug operators to venture far offshore. The vessels are often not large enough and the crew may not be sufficiently trained to effect distant ocean recovery in heavy sea conditions. Because the environmental impact of far-offshore casualties is minimal (or at least not obvious), the attendant publicity and public as well as governmental interest in such incidents is limited.

The effectiveness of rescue towing varies by region in the United States. The East Coast is densely and rather uniformly punctuated by major port facilities, from Portland, Maine, to Miami, Florida, a distance of some 1,250 nautical miles. The maximum distance between ports having a rescue towing capacity is 325 miles. Seagoing tugs of 4,300 shaft horsepower (SHP) and under are usually available throughout that range. The majority of coastwise tows rarely exceed 60 hours. Tugs exceeding 5,700 SHP with a bollard pull of more than 120,000 pounds are less common, but often available. There is one salvage-equipped vessel based in New York that is available for salvage and pollution response when not engaged in long-distance coastal towing. It is 210 feet long, with a bollard pull of 180,000 pounds, and is all-ocean operational with a top speed of 15 knots. Coastal towing occupies perhaps 30 to 40 percent of its time. In sum, the East Coast is fairly well covered in terms of conventional tugs able to respond to emergencies. It was reported to the committee that a tug can be on scene, ready to assist, within 12 hours anywhere on the East Coast (Sweeney, 1993).

Of the three coasts, the Gulf Coast has perhaps the largest array of rescue-capable vessels. The high-horsepower tugs and supply vessels that support the offshore oil industry often have ample accommodations for salvage crews, good towing capability, and large after-deck working areas, and they are generally available, subject to existing oil field contracts. High-horsepower tugs and tug-supply vessels of up to 8,000 to 10,000 SHP are usually available, if not abundant. Most equipment in the Gulf is under contract to oil companies and oil-field-related contractors with significant standby time; a short-term release from these contracts usually can be secured for emergency purposes. The public expectation of successful salvage in environmentally sensitive areas has motivated significant marine industry cooperation in emergencies, particularly by oil-related companies. The west coast of Florida and the western end of the Gulf are not as well covered as the central portion, due to the limited offshore oil and gas industry in those areas. These areas are dependent for response on the tugs involved in the normal coastwise barge trade. The Florida Straits, at the eastern end of the Gulf, is of specific concern due to the area's environmental sensitivity. The Coast Guard is studying the potential for vessel routing schemes and/or exclusionary zones to provide additional protection in the event of a vessel casualty.

On the West Coast, which has fewer major ports than either the East or Gulf coasts, the long distances between ports mean that fewer tugs are available in the intervals. For example, the distance between San Francisco, California, and Portland, Oregon, is approximately 600 nautical miles, which may mean two days running for an ocean tug. The major tugboat companies have fleets as modern as in the rest of the country, perhaps smaller in numbers but with capabilities ranging to 9,000 SHP, which is not found regularly on the East Coast. The availability of rescue towing from Seattle, Washington, north to Alaska is a concern due to the remoteness of the area and the limited tug and barge activity. The only two U.S.-owned, dedicated salvage vessels are stationed on the West Coast, one in Astoria, Oregon, the other in Seattle,

Washington. Both are primarily equipped for strandings and are excellent salvage platforms. However, low horsepower and bollard pull limit their effectiveness as prime offshore rescue towing vessels. The Valdez, Alaska, to Long Beach, California, oil run has been the focus of considerable attention since the *Exxon Valdez* accident. Oil companies and coastal states are studying rescue and escort towing. One study focusing on tugs and escort vessel capability to assist a disabled tanker in Prince William Sound, Alaska, concludes that existing vessels—about 5,700 SHP—are adequate, but two may be required under certain circumstances (Disabled Tanker Towing Study Group, 1993).

The long distances between ports on the West Coast has focused attention on options such as vessel routing schemes and exclusionary zones, such as those being investigated for the Florida Straits. There are few locations in the gaps between the ports to be of additional towing capability, and the extreme distances would require multiple basing—an extraordinary expense—to achieve coverage. As a result of the physical location limitations and financial impact, changes in voyage practices are being investigated.

The relationship between size and a tug's ability to control or tow a vessel is not well documented. A study was conducted for the committee (Appendix I) to assess the adequacy of tug and towing capability on U.S. coasts. The study used a relatively simple procedure to analyze the size of the tug required to tow various size tankers. The analysis consisted of calculating the mean wind and wave forces on the tanker and tug and converting them into required bollard pull of the tug. Various wind and wave combinations were addressed, and tug horsepower requirements estimated. The analysis showed that 7,000-horsepower tugs can handle both a very large crude carrier (VLCC) and a 140,000-deadweight-tons (DWT) tanker up to just short of a 20-foot (6-meter) sea state and for handling an 80,000-DWT tanker up to a sea state of 21.5 feet (6.5 meters). Beyond those points, a larger tug, or two tugs, would be required. The study additionally addressed the joint probability of 20- to 21.5-foot (6- to 6.5-meter) seas and an onshore wind to predict the most serious scenario, i.e., when conditions would be beyond the capability of a single 7,000-horsepower tug. The percentage of time that these two conditions occur jointly varies from a low of 0.3 percent of the time on the East Coast to a high of 1.4 percent off Alaska. In lesser seas, tugs as small as 3,500 horsepower were shown to have sufficient power to provide for control under certain wave and wind conditions.

Two important constraints limit the overall reliability and effectiveness of rescue towing in the United States. The first is that virtually the entire U.S. rescue towing fleet, with rare exceptions, is engaged primarily in towing vessels from point to point on a commercial basis. This practice may limit the choice of available equipment for emergency salvage response. The second limiting factor is the size and quality of tug and tug-supply boat crews. A typical navigating crew on a U.S. tug trading coastwise is five to seven persons, with few opportunities to gain experience in rescue towing under adverse conditions. This is not a criticism but a fact. Tug crews are trained to tow in sheltered waters, not to connect to distressed vessels in adverse conditions. Additionally, the lack of suitable towing systems aboard most vessels (discussed under Salvage Readiness of Vessel and Crew, page 53) makes it difficult to attach a tug in heavy weather.

In light of the public interest in marine environmental protection, there have been calls for industry or public support of salvage, particularly rescue towing, to ensure available support in the event of a major casualty. This idea derives from the regulatory consequences of OPA 90, which mandate substantial standby resources for pollution response in every significant coastline location of the United States.

The availability of dedicated salvage vessels, placed strategically around the United States, to provide salvage support and rescue towing capability is the subject of considerable discussion. As noted earlier, the cost of dedicated salvage vessels is very high, and impractical from a commercial standpoint. The United States allowed its subsidies through the Navy to wither away because of the expense and the minimal offsetting commercial salvage revenue (see Appendix F). France, South Africa, and Spain continue to subsidize salvage capability by using public funds to pay for some operating expenses. France and South Africa have joint ventures with salvage vessel operators, which permit the vessels to be employed commercially. The French arrangement calls for the government to share 50 percent of the commercial salvage awards; the government's share has not risen above 14 percent of the annual operating costs, and less than 5 percent is more typical. The committee did not determine the feasibility or viability of a dedicated rescue towing capability covering the entire United States.[6] It should be noted, however, that the U.S. coastwise towing industry is the largest in the world, and nondedicated tugs are more readily available for rescue towing, subject to regional differences, than in those countries that have subsidized a dedicated towing capability.

Some observers suggest that the limited number of towing vessels over 6,000 SHP leaves the United States unprotected for handling some casualties. While U.S. coasts have been protected adequately in the past, the growth in vessel size, particularly oil tankers, has generated concern. Recent studies have demonstrated that vessels of 6,000 SHP may not provide effective rescue towing for larger vessels. The study conducted for the committee suggests that the capability of available tugs in the 6,000- to 7,000-SHP range is, however, adequate to provide for control of a VLCC in all but the worst weather conditions. These findings are reinforced by the Prince William Sound towing study, although, as noted earlier, that study also found that two tugs of that size may be needed to control a fully laden VLCC (Disabled Tanker Towing Study Group, 1993). Future parts of the Prince William Sound study will include simulations that are expected to shed additional light on the subject of towing disabled tankers.

Findings

The control and management of a disabled vessel, although perhaps not rescue towing per se, can be accomplished with tug assets that exist throughout the United States. Tugs of 6,000 SHP and higher, required to control or manage large vessels such as VLCCs, are available throughout the nation but are limited in number. The ready availability of tugs is not continuous along the coastline, particularly on the Pacific Coast with its long distances between ports.

Tugs that could be used for control and management of distressed vessels are not designed for at-sea rescue towing, and their crews normally are not trained for these activities. Vessels that may require towing generally are not equipped with emergency towing packages, although international recommendations concerning emergency towing have been advanced since the early 1980s. There are moves, however, both in the United States and internationally through the International Maritime Organization (IMO), to make such recommendations mandatory.

Traffic routing schemes and exclusionary zones may be used to minimize risk and prevent accidents, further reducing the need for rescue tugs.

MARINE FIREFIGHTING

Fighting fires aboard vessels is complex—it is complicated by weather, vessel type, fire type, cargo type, accessibility, and damage stability. Success depends on the ability of the salvor or firefighter to mobilize and utilize limited resources. But while

[6] The committee received information that suggests that the costs of operating such a vessel would be about $5,000 per day, not including debt service on the capital cost of the vessel ($16 to $20 million) or the cost of onboard salvage personnel (White, 1993).

the marine firefighting environment is multifaceted, the manner of extinguishing fires is fairly straightforward. The four basic processes are heat extraction, fuel isolation, oxygen dilution, and chemical inhibition of reactions: the most commonly used approaches are heat extraction and fuel isolation (NRC, 1986).

1. *Heat Extraction.* Water is the primary firefighting agent used for heat extraction and cooling action for the protection of firefighters. Professional marine firefighters contend that it is not the volume of water that is critical but its strategic placement. In fact, overapplication of water has resulted in the sinking of several ships. Marine firefighters estimate that the cooling load required to contain and control a fire is on the order of 0.15 gallons of water per minute (GPM) per square foot of deck area (Williams, 1993). The volume of water required by a fire like the one on the *Mega Borg* (roughly 30,000 square feet of deck area) requires a pumping capability of roughly 4,500 GPM. As such, portable pumps of around 2,000 GPM are effective. A sufficient supply of such pumps is available around the country, and they are efficient to transport from storage to casualty sites. In some cases, the Coast Guard has assisted with transportation.

   While there are not significant numbers of dedicated fireboats in the United States, the number of tugs and workboats equipped with enhanced firefighting capability has increased. This capability is particularly prevalent on the Gulf Coast, where approximately 14 offshore support vessels are equipped with fire pumps rated at 5,000 GPM or greater. In addition, it is not uncommon for harbor tugs to have some enhanced firefighting capability. Several U.S. ports have dedicated fireboats, some of which are crewed and equipped to fight offshore fires. For example, in the recent Tampa, Florida, casualty, the local fireboat crew was trained by a specialist in marine firefighting, and they were able to control the fire with advice via radio from the marine firefighting specialist.
2. *Fuel Isolation.* Water alone may not be sufficient to fight marine fires. Foam often is used to smother liquid-fueled fires. It is available around the country through marine firefighting companies, foam manufacturers, refinery stocks, municipal fire departments, and the military. These resources must be managed to ensure that the required foam can be obtained quickly if required. Availability of foam and efficient delivery to the casualty site have improved since 1982.
3. *Chemical Inhibition.* Chemical inhibition of reactions is accomplished through the use of dry chemicals and halon gas. This is generally most effective when applied before the fire gets too hot, as there is no cooling effect or protection from a reflash associated with the use of these agents. Unfortunately, the use of halon gas has some negative environmental effects, and it is being phased out as a firefighting agent. No equivalent substitute has been found to date, and new ships are being built using dangerous and less-effective carbon dioxide in their fixed firefighting systems. This situation has resulted in an increased interest in more advanced water fog systems. Emerging technology shows promise for the delivery of dry chemicals in encapsulated form using foam as a medium.
4. *Oxygen Dilution.* Oxygen dilution involves blanketing spaces with an inert gas, usually carbon dioxide. The viability of this approach depends on whether the space can be sealed effectively, and whether a sufficient volume of carbon dioxide is available. Carbon dioxide in bulk is the most commonly used smothering agent. It can be dangerous, however, and there have been many fatal accidents involved with the use of carbon dioxide as a firefighting agent,

some involving actual shipboard fires, others involving accidental releases, usually in shipyards.

Marine firefighters need basic marine firefighting experience, correctly sized equipment, appreciation for wind and sea effects on a vessel, and understanding of vessel construction and stability. Few companies specialize in marine firefighting in the United States, or, for that matter, in the world. They are supplemented by salvage companies that offer firefighting expertise as part of their overall package, and normally call in specialists for serious fires. While not great in number, specialized U.S. marine firefighters have effectively handled marine fires over the past ten years. Increased attention to fire prevention and improved onboard firefighting have led to fewer significant casualties. Thus, recent experience indicates the nucleus of highly trained marine firefighting professionals appears adequate to meet the need. Furthermore, over the last decade, U.S. marine firefighters have developed and refined firefighting systems and methods. Marine firefighting equipment is "forward deployed" in depots on each coast. Additional equipment is packaged for fly-away delivery, in which experts and equipment are flown to a casualty and utilize platforms of opportunity.

There are, however, few opportunities to train and gain on-the-job experience. More than 30 U.S. training schools advertise a marine firefighting curriculum. These curricula focus on shipboard firefighting, rather than the firefighting methods practiced by salvors in a serious fire using off-vessel firefighting assets and approaches. Only five or six meet the Coast Guard licensing requirements for basic and advanced firefighting by providing significant hands-on training in addition to classroom instruction. The remaining schools either are primarily classroom-based or are designed to fill a niche, such as barge firefighting, use of portable and semiportable equipment, and cadet training. The Coast Guard and the Navy operate schools for firefighting and damage control at various locations around the country for their personnel.

The recent track record of marine firefighting has been good. The industry has demonstrated the ability to deliver personnel, foam, and portable equipment to fires in a timely manner. In the *Jupiter* incident in the Great Lakes, a marine firefighter was on the scene within 12 hours after notification, and the fire was extinguished within two hours after arrival. The success of such efforts has depended on the availability of vessels of opportunity. To date, finding such vessels has not been a major problem. In the *Mega Borg* fire, 8 vessels, 6 with firefighting capability, were on the scene within 36 hours; in the *Jupiter* fire, the Coast Guard provided the platform from which the fire was extinguished. The number of boats with enhanced firefighting capability has increased since 1982, although proximity to the casualty and the availability of vessels of opportunity when needed are always uncertain. Availability of foam has improved.

Marine firefighting readiness in the United States is improving rapidly, due to two important new mandates:

1. New requirements for escort tugs equipped with some firefighting capability in some of the major port areas.
2. OPA 90 salvage planning requirements, which have resulted in an increase in numbers of prepositioned, strategically located firefighting systems.

Overall, significantly more resources are available today for responding to a vessel fire than were available in 1982.

Still, several concerns remain. These are usually of a local nature and would apply to specific ports. In some port areas, local firefighters have little or no training in marine firefighting. Not all ports have dedicated fireboats and marine firefighters. The availability of tugs with firefighting monitors and foam varies considerably around the United States. While the marine industry has shown that highly specialized equipment can be deployed quickly to a casualty site, the response time requirements change considerably when fire threatens life, as would be the situation with a passenger vessel casualty, for example.

Findings

The marine firefighting capability of the United States has improved significantly in the years since 1982. The development of prepositioned, fly-away marine firefighting systems, continuing modification of existing tugs to increase firefighting capability, and the addition of escort tugs with designed-in firefighting systems in some ports has improved the overall marine firefighting posture. Additional firefighting equipment is being deployed at coastal sites as a result of actions taken by salvage companies and the marine transportation industry to comply with salvage readiness regulations stemming from OPA 90.

Gaps still exist in marine firefighting experience and capability in certain port areas with respect to equipment, particularly a lack of waterborne firefighting capability (fireboats or tugs with fire monitors) and in some instances, insufficient training of local fire department personnel to respond to shipboard fires. These deficiencies have been successfully countered to date by the rapid transport of equipment and marine firefighting experts to serious fires.

CARGO TRANSFER

Refloating a grounded tanker or dry cargo ship often requires the shifting of weight, usually by cargo transfer. In fact, this will always be the case unless the vessel grounded at low speed and can be pulled free, or a rising tide can be used to advantage. In some cases, this can be done by internal cargo transfer; in others, cargo must be transferred to a barge or other receiving vessel. In extreme cases, jettisoning may be an option.

Internal cargo transfer may be effective if a modest change in trim[7] is sufficient to refloat the vessel. External cargo transfer is required if a more radical change in trim or draft is necessary. Internal transfer often can be conducted with liquid cargoes without outside assistance, while external transfer requires other ships or barges and ship-to-ship lightering equipment. Weights should never be transferred, either internally or externally, until the effects on trim, stability, and stress have been determined.

If the stranded or damaged vessel is a loaded tanker, it will probably be necessary to transfer cargo having a low flashpoint. If at all possible, such a transfer should be made under inert conditions. This generally is not a problem in a ship-to-ship transfer; but in internal operations involving, for example, transfer of cargo to a dedicated ballast tank, it may be necessary to improvise to introduce inert gas into the receiving tank. If the shipboard inert gas plant is inoperative following a casualty, then it may be possible to use portable inert gas generators.

Internal Liquid Cargo Transfer

Internal transfer, when possible, is preferable to external transfer because it usually can be accomplished in a timely manner, not being dependent on outside equipment that must be brought to the scene. The goal is to alter the trim and refloat

[7] Trim refers to the horizontal position of the vessel, or the difference between forward and aft draft. Draft refers to the depth of the vessel in the water.

the vessel. Possible alternatives should be reviewed; there may be several, and, in accordance with safety considerations, the least complicated method should be chosen. On a tanker, there may be a choice of transferring cargo, fuel, or water into empty cargo tanks, segregated ballast tanks, or void spaces. The greatest changes in trim can be achieved by transferring cargo to or from tanks at the ends of the ship, although in certain cases transverse transfer to induce listing may help.

Transfer of cargo into empty cargo spaces is generally uncomplicated, provided the cargo piping system is intact. The transfer may be effected by gravity alone or by the use of cargo pumps. Transfer of cargo into segregated cargo spaces may be difficult on vessels without modifications that bypass systems designed to passively prevent the passage of oil into ballast spaces.[8]

Transferring cargo into segregated ballast spaces can be very effective because large quantities can be moved in a short time from the extremities of the cargo spaces, thus having the maximum effect on trim. This is best accomplished when connections between the cargo and ballast piping systems can be made in the pump room using portable spool pieces. If a ballast manifold exists, cargo may be transferred by rigging a hose between the cargo and ballast connections. The use of void spaces to receive cargo is the least desirable of the internal transfer options. Void spaces are generally small, and piping arrangements are not easily adapted to cargo transfer.

While fixed systems to facilitate this transfer are not permitted under IMO guidelines (to prevent contamination of ballast spaces), preplanning and the design and availability of portable equipment for the particular vessel to facilitate the process would allow the utilization of this technique under extraordinary emergency situations. Simple vessel modifications can facilitate the transfer process. A major West Coast tanker owner described to the committee a plan to modify its vessels to enable the transfer of cargo to empty spaces or to segregated ballast tanks in the event of an emergency (Stiehl, 1993).

External Liquid Cargo Transfer

The most commonly used technique for refloating a stranded ship is to lighten her by removing some of the cargo. It is a method that works in all but the most unusual circumstances, such as those involving massive damage to the ship. In the 1970s, the major oil companies developed ship-to-ship transfer techniques lightering operations to enable them to achieve economies of scale offered by VLCCs on voyages from which they would be precluded, being too large to enter the delivery port. Because of this, a large amount of ship-to-ship transfer equipment is available, together with an adequate pool of experienced personnel familiar with this operation.

Transferring cargo between ships, as a normal operation, requires only large fenders, cargo hoses, standby pollution control equipment, and people familiar with berthing of vessels and lightering operations. The situation changes for stranded or damaged vessels. If the ship has lost power, an external power source for the transfer pumps is required. Hydraulic transfer pumps with diesel power packs are readily available on a fly-away basis and can be delivered quickly anywhere in the United States.

Lightering a stranded vessel may be complicated if she is grounded where it is difficult for another vessel to approach without stranding. This problem sometimes

[8] The unified interpretation of MARPOL Regulation 1(17) has a provision for emergency discharge of segregated ballast by means of a connection to a cargo pump through a portable spool piece. The same regulation requires nonreturn valves to be fitted on segregated ballast connections to prevent passage of oil to the segregated ballast tanks. This regulation thus seriously hampers the emergency transfer of cargo into empty ballast spaces in the event of a vessel casualty.

can be solved by using small lightering vessels or barges. It may also be possible to use floating hoses to transfer cargo from a stranded vessel to a point farther offshore, where an offtaker vessel can approach safely.

Nonliquid Cargo Transfer

From a salvage standpoint, nonliquid cargoes are in many ways more difficult to handle than liquid cargoes. In addition, there are significant concerns regarding the increasing volume and variety of hazardous cargoes (see Salvage in the Presence of Hazardous Cargo, page 48). Nonliquid cargo generally must be offloaded to refloat the vessel; little capability or space exists aboard dry cargo vessels to transfer cargo internally. The jettisoning option is impeded by the same obstacles that block jettison of liquid cargoes (see Jettisoning, page 49). Offloading may require a major mobilization to obtain resources and labor for moving bulk cargo.

General Cargo Transfer

Weather is critical in ship-to-ship transfers. Under most circumstances, cargo transfer is a routine operation. Even when it has to be performed under adverse conditions, such as after grounding or a severe casualty, the resources to handle the additional problems are available and generally can achieve a satisfactory outcome. Oil companies have developed sophisticated procedures to minimize risk in cargo transfer, and liquid lightering operations have a safety record that is comparable to shoreside operations (USCG, 1993).

The availability of suitable vessels (particularly tank vessels to receive liquid cargoes when a casualty requires lightering) is generally a minor concern. In the case of the *Exxon Valdez*, empty vessels coming into the Port of Valdez were readily available, and lightering proceeded in an expeditious manner. On the other hand, in the case of the *Minerva* in the Gulf of Mexico, a barge suitable for offloading bulk cargo was not available in a timely fashion, due to previous commitments. In that case, it took almost a week to obtain the barge. The situation was not time-critical, but it does indicate the uncertainty of obtaining suitable receiving vessels and equipment.

In most of the United States, the availability of lightering vessels is not a problem. In investigating the response to tanker casualties in three locations on the East Coast, the committee found that adequate numbers of suitable lightering vessels were available on-scene, with supporting tugs, within six hours of notification. Response times vary considerably by coastal region, depending on the remoteness of the accident scene and the amount of vessel traffic. Lightering barges prepositioned by pollution response companies over the past two years represent a substantial increase in dedicated resources for lightering since 1982. Tank vessel owners are required to establish a lightering plan as part of their contingency plans under OPA 90.

Findings

Internal and external transfer of liquid cargo is not generally a problem, due to the tanker industry's extensive experience with ship-to-ship transfer over the years, and the development of written procedures arising from that experience. Ease of lightering depends on how close the situation is to a normal ship-to-ship transfer. Considerable ship-to-ship transfer equipment is available, either within port areas or on a fly-away basis.

The major constraint on external transfer of liquid cargoes, or lightering, is the availability of suitable vessels to receive the cargo. Suitable lightering vessels appear to be available throughout the United States. Committee investigations have shown that resources for handling liquid cargoes can be on-site in most coastal areas within 12 hours. For nonliquid cargoes, the constraining resource is at-sea cargo-handling equipment.

DAMAGE STABILITY INFORMATION

The development of personal computers and sophisticated software has added a very powerful weapon to the salvor's arsenal. Salvors once operated by the seat of their pants when deciding what they could or could not do in their efforts to stabilize or refloat a damaged ship, particularly with regard to the possible effects of an already damaged ship. The calculations necessary to provide relatively accurate engineering data were too complex and time consuming to allow for numerous what-if solutions. In addition, the basic data required to make those calculations often were not available.

Salvors require much of the same information used by naval architects to design a ship. The original computer program used to calculate hydrostatics, floodable length, intact and damage stability, and longitudinal strength was the Ship Hull Characteristics Program (SHCP) developed for the U.S. Navy in the mid-1960s. This program, which ran on mainframe computers, calculated the hydrostatic properties of a freely floating ship in either intact or damaged condition. While the SHCP was intended to be a design tool, in the 1970s and early 1980s it was one of the main computational tools available for salvage situations. SHCP was an excellent design tool, but it was never intended to be a salvage program. Data entry for hydrostatics and damage stability was time-consuming; it could take several person-days to enter the information for one ship. Once the data was entered, it could take an hour to run a single damage stability case on the early mainframe computers. Unlike a modern dedicated salvage program, SHCP did not allow input of cargo weight by tank or hold. Rather, the user had to calculate weight and centers of gravity for the cargo and combine them with light-ship weight and centers to create a weight distribution curve. Use of SHCP for salvage analysis was further limited because the program could not directly handle all the calculations needed in a grounding situation, nor could it provide guidance for hydrostatically balancing a damaged tank.

In the early 1980s, analytical and computer programs were developed that reasonably approximated ship characteristics with only minimal data input (Porricelli, Boyd, and Schlieffer, 1983). Programs based on this methodology were related closely to the simple calculations traditionally performed by salvors to estimate ground force reaction, stability, and other variables. The characteristics of many ships were used to develop these programs, so the statistical analysis led to very reasonable estimates of vessel characteristics without requiring large amounts of data input. One of the key forces driving the development of salvage programs in the 1980s was an explosion in the use and availability of personal computers with sufficient power to handle damage stability problems.

It is safe to say that every shipping company had its own approach to handling salvage. In the mid-1980s, some shipowners began to search for ways to improve their salvage response capabilities. Some pre-entered the intact data for their ships into general naval architectural programs, with the idea of saving time in the event of an accident. Other owners created salvage manuals containing the results of many damage stability calculations, to provide data for quick reference in case of an incident.

In 1985, for example, one major oil company wrote several programs intended to be used in conjunction with SHCP for rapid development of loading conditions for analysis of damage stability and longitudinal strength. These programs, combined with pre-entered data, provided excellent salvage tools. With these programs and data loaded into laptop computers, the company's naval architects were prepared to fly to the scene of an incident and respond immediately.

Since 1986, many computer programs for salvage assessment have been developed and are commercially available at reasonable prices. In the United States, there are at least a dozen programs capable of supplying damage assessment information.[9] Thus, there were few analytical tools available to salvors into the mid-1980s, whereas there is now a wide choice. The U.S. Coast Guard, the Navy, most major oil companies, and many of the larger shipping lines maintain in-house capabilities using one of the newer computer programs. For shipowners who lack the technical staff or who do not wish to maintain in-house salvage programs, classification societies will provide this service. With a program and preloaded ship characteristics loaded on a portable computer, the salvor can quickly estimate damage stability, residual strength of the damaged structure, oil outflow, and grounding force. The ability to run sample loading conditions in minutes allows the salvor to rapidly calculate the effects of proposed cargo shifts on the vessel.

Today's salvor has tools that take much of the guesswork out of salvage operations. Fairly precise calculations can help determine what to do to refloat the ship without inflicting damage, while allowing for damage conditions. These tools, combined with the experience of a competent salvor, go a long way toward ensuring successful salvage. However, as with all computer programs, the user must understand the assumptions and limitations of any program and use sound judgment in applying the results. Too often, as the committee was reminded, the results of a computer program are treated as though they are "good as gold," when in fact they may be based on erroneous assumptions or limited data and may yield answers that are not viable in a given incident (Edgar, 1993). The salvor still, as was the case prior to these computer aids, makes significant estimates of the vessel's damage. These estimates are vital to the solution, whether computer-aided or arrived at manually.

In the past, salvors had limited access to damage stability information and tools for working with that information. This situation has improved, not only in terms of the tools available but also in the percentage of the marine fleet that maintains damage stability data, either on the vessel or ashore within easy reach. In addition, OPA 90 requires that all tank vessels have on board or readily available information on damage stability and vessel construction.[10] The classification societies and the Coast Guard maintain vast databases of stability and construction data that is becoming available on a real-time basis for use in a marine casualty.

[9] The following list is representative of the programs available to assist in salvage analysis.

Acronym	Descriptor	Organization
SHCP	Ship Hull Characteristics Program	NAVSEA
SCPS	Shipboard Contingency Planning System	ABS
GHS	General Hydrostatics System	Creative Systems
FCCS	Flooding Casualty Control Software	NAVSEA/USCG
POSSE	Program of Ship Salvage Engineering	SUPSALV/Herbert Engineering Corp.
CARGOMAX	Cargo Loading Program	Herbert Engineering Corp.
HECSALV	Herbert Engineering Co. Salvage Program	Herbert Engineering Corp.
AUTOHYDRO	Automated Hydrostatic Program	Coastdesign
BALLAST	Balanced Loading by Automated Stability & Trim	Pacer Systems
NAUTILUS	Nautical Loading System	New Wave Systems
PLAN	Programmed Loading & Analysis	Frank Chou Associates
MECAS	Marine Engineering Casualty Analysis System	Frank Chou Associates

[10] 33 CFR 155.1035(c)(5).

Findings

The availability of damage stability information, either on a vessel or readily accessible on shore, has improved considerably in the past decade. This trend will continue, especially for tankers calling on the United States, due to the OPA 90 requirement for ready availability of such information.

The number and capabilities of tools available to salvors have increased dramatically with the proliferation of computer programs that can provide more detailed information in less time than was possible in 1982. However, these programs depend on the assumptions inherent in computer models, the completeness and accuracy of the data, and the analysis of the results. Salvors need to recognize the limitations of such tools.

SALVAGE IN THE PRESENCE OF HAZARDOUS CARGO

In recent years, hazardous cargo shipments have become more common on the high seas. This trend is commanding serious attention from mariners, environmentalists, government authorities, and salvors. The carriage of liquefied natural gas, arsenic, ammonia, chlorine, cyanide, and pesticides, either in containers or in bulk, poses a serious potential hazard.

It is certainly a credit to the carriers of such products that accidents have been few, but there have been some incidents. Some arsenic containers were recently lost off the coast of New Jersey; a lash barge (i.e., a lighter stowed aboard a ship) that broke loose from an ocean carrier during heavy weather was delivered to New York emitting deadly fumes caused by the commingling of cyanide and acids. In both cases, full encapsulation of the lost cargo was required. In the arsenic case, underwater robotics played a major role in the recovery. The lash barge was emptied successfully and salved after careful separation of the cargo by a team of highly trained hazardous materials (hazmat) specialists.

Salving hazardous cargoes begins with identification and analysis of the risk. This task may require sophisticated testing by highly trained, specialized chemists or technicians; most salvors lack this capability. A study by the Chemical Manufacturers Association (CMA) concluded that there is no capability in the United States to respond to a major spill of hazardous materials (CMA, 1992). A high degree of specialization of skills and equipment is usually required, not only for onboard salvage, but also throughout the materials handling process to the final destination or disposal.

Salvors in the United States must rely on hazardous waste remediation companies, which are subcontracted to identify, handle, and dispose of hazardous and toxic cargoes. Only one U.S. salvage company, based in the Northeast, has in-house capability for remediation and disposal of hazardous and toxic cargoes. This capability was developed through acquisition and is not the result of training or experience within the company, but it has allowed for the integration of salvors and specialized hazmat personnel in cross-training and cooperative salvage solutions.

Handling and disposal of dangerous or hazardous cargoes requires not only personnel trained to identify and quantify the risk but also specialized equipment for handling and disposing of such cargo. As a minimum, hazmat personnel are required by law to have 40 hours of training, approved by the Occupational Safety and Health Administration, with an eight-hour refresher course each year. Hazmat personnel also must have, by law in most states, a complete medical and physical examination every six months. By contrast, U.S. salvage contractors seldom have the requisite materials-handling equipment, analysis capability, environmental monitoring tools, appropriate transportation equipment, or safe disposal techniques. It appears that the only solution available to salvors today, with rare exception, is preselection of an experienced hazmat contractor who can provide support as needed. Yet there is little evidence of routine cooperation among salvors and hazmat contractors today.

Regional and state emergency services teams generally lack the knowledge and experience to board a vessel and deal with dangerous cargo. Moreover, they often are precluded by their regulatory role from conducting such operations. Recently, Coast Guard strike teams have undergone extensive training in the assessment and handling of hazardous cargoes. The strike teams are probably the most experienced and trained forces available to respond to a hazardous material casualty, although to date their role has been limited. However, if the gap in commercial capabilities persists, the strike teams may serve as the primary resource in such incidents.

The salvage role of the government, particularly the strike teams, continues to evolve. Where government enforcement and response should begin and commercial salvage should join and complete the effort continue to be debated. Casualties involving hazardous cargoes pose the highest likelihood of danger to the public and the environment. The federal government is expected, at minimum, to continue to maintain a significant presence to guide salvage efforts to successful conclusions. Ultimately, the effectiveness of that presence will depend on the quality of commercial salvage capabilities.

There does not appear to be sufficient economic incentive for the salvage community to develop a comprehensive, self-sufficient, total response capability for hazardous cargo casualties. The most viable alternative, employed only to a limited degree, is for the salvor to have a preexisting subcontract with a qualified hazmat contractor. It is only by preselection and integrated training *before* a casualty occurs that a successful result can be assured. Assessment of the nation's ability to handle hazardous and toxic cargo properly may well be the next principal focus of government, environmental, and marine industry interests in the 1990s, now that OPA 90 has been implemented.

Findings

The marine salvage community does not normally maintain in-house capabilities to perform salvage in the presence of hazardous substances. Most salvage companies must obtain the support of an experienced, land-based hazmat contractor to proceed with salvage where hazardous materials are involved. With rare exceptions, little experience has been gained with this type of combined effort.

Coast Guard strike teams have extensive training in responding to marine incidents involving hazardous materials, although their role in commercial casualties is not well defined.

JETTISONING

The possible need to discharge oil to save a ship is recognized in the International Convention for the Prevention of Pollution from Ships (MARPOL 73/78). MARPOL is in force in 78 nations, including the United States, yet U.S. statutes impose penalties or liability for oil discharges into the sea that otherwise would be permitted by the convention. The committee convened a symposium to assess the issues involved in the intentional discharge of oil during salvage operations as part of a study of U.S. salvage capability.

The Symposium on the Purposeful Jettison of Petroleum Cargo, held February 23, 1993, addressed the need to clarify U.S. law concerning intentional discharges of cargoes to save ships and prevent loss of larger amounts of cargo, and the implications of advances in oil spill contingency planning, environmental data acquisition, and spill trajectory forecasting. Speakers addressed the historical context of jettison, environmental monitoring requirements, spill trajectory modeling technology, and the legal status of jettisoning under federal and state laws and international treaties. These presentations were followed by two panel discussions that focused on the jettisoning of oil as one option for response to an accident scenario. The report and recommendations on the purposeful jettison of petroleum cargo are attached in Appendix B (the proceedings of the symposium are published along with the report and recommendations, in a separate publication [NRC, 1994]).

A number of themes and findings emerged from the symposium presentations, based on the committee's analysis. Speakers generally agreed that jettisoning of oil can be a valuable salvage tool and should be considered as an option, but one to be undertaken only when failure to take such action would likely result in loss of the vessel and release of the entire cargo. A deliberate discharge of a small volume of oil may be the only practical alternative in certain time-critical situations. Conventional alternatives such as lightering may prove impossible due to the timely availability of appropriate assisting vessels.

Jettisoning has been rare in recent years, and contemplation of its use may be even more restricted by OPA 90, which introduced a new strict liability standard for damage from oil spills and established criminal sanctions for spillers. The symposium speakers' varying interpretations of OPA 90 reflect the ambiguities in federal and state oil pollution laws and confusion within the maritime community concerning the legal status of jettisoning (NRC, 1994). The Congress did not consider implications for salvage in enacting OPA 90, and the resulting uncertainty over liability clearly is a factor in the reluctance to jettison. Furthermore, an intentional discharge would violate the Federal Water Pollution Control Act (FWPCA), and state laws may impose additional liabilities.

The direct way to protect salvors and other responders from liability may be to amend the National Contingency Plan (NCP) to clarify the procedure for deciding to jettison and to place the responsibility for authorizing such action solely on the federal on-scene coordinator (FOSC). As a practical matter, such an approach could obviate the need for Congress to amend OPA 90 or to await a judicial interpretation following an incident of jettisoning. This change would not entirely solve the problem, however, because OPA 90 does not preempt state law, and the salvor may be exposed to additional direct or indirect liability under general maritime law or various state laws.[11] In any case, clarification of oil pollution laws undoubtedly will require further judicial or regulatory interpretations.

Participants in the symposium panel discussions differed as to whether jettison would be an appropriate response to the given accident scenario. This disagreement demonstrates the difficulty and subjective nature of such decisions and suggests a need for standard, objective decision-making criteria. Such criteria could help expedite a process that inevitably involves multiple decision makers and special interests. Eight criteria were suggested as fundamental conditions that must exist before any oil is jettisoned:

1. Time pressures demand immediate action.
2. Deliberate discharge of the proposed amount of oil is likely to save the ship and the remaining cargo.
3. All other salvage options, such as internal cargo transfer and lightering, have been exhausted or considered and rejected.
4. Failure to jettison is likely to lead to loss of the ship and release of the remaining cargo. The principal issue is likely to be whether the ship will break

[11] States have no jurisdiction (for this purpose) beyond the 3-mile territorial sea. States do assess penalties and attach liability, however, to those responsible for spills that originate outside the territorial sea and drift into it.

up in bad weather, so information is needed concerning tides, currents, and approaching storms.

5. The condition of the stranded vessel is such that she could be refloated following the jettison, and the remaining cargo saved.
6. All necessary preparations have been made, including the marshaling of tugs, if available, to refloat the ship quickly after the discharge.
7. The FOSC is monitoring the situation continuously to ensure that jettisoning remains the only viable option.
8. Preparations are underway to clean up the discharged oil. Information is needed concerning spill trajectory, characteristics of the oil, physical environmental conditions, containment and recovery measures, geology of the impact zones, toxicological sensitivity of vulnerable species, and ecological characteristics of vulnerable areas.

Finally, two general factors that may impede sound salvage practices were mentioned. Several speakers indicated that, even when jettisoning appears to be the correct technical decision, the FOSC in the decision-making exercise only recommends this action to superiors first (the district commander and ultimately the commandant). This places the issue in the political arena, as occurred in the *Argo Merchant* case. Under these circumstances, and without specific criteria on which to base a decision to jettison, public environmental concerns effectively may block action.

Another issue is the uncertain legality of discharges that may occur during the normal course of salvage. A number of practices of salvors could be deemed a form of jettisoning, as they may result in a discernible discharge of oil. Examples include pumping out a flooded engine room, pressing down of dirty ballast tanks, expelling water from a flooded cargo or fuel tank, using compressed air to press out damaged tanks, displacing oily water with buoyant material, and operating skimmers (which, in separating oil and water, may discharge small quantities of oil). Such actions result in minimal pollution and likely would be part of an approved plan of action; yet, regardless of their benefit, these incidental discharges may violate the FWPCA.

Findings

Jettisoning of petroleum cargo can be a valuable salvage tool, but should be considered as an option to be undertaken only when failure to take such action might and probably would result in loss of the stranded vessel and release of the entire cargo. However, a number of unresolved issues tend to inhibit the reasoned use of jettisoning.

In the committee's judgment, these issues should be resolved before a marine casualty occurs where a deliberate discharge of oil may be warranted. Otherwise, in the absence of legal certainty, a salvor may reject the jettison option arbitrarily, even when it may be the only means available to avoid a catastrophic spill.

Marine pollution response increasingly is being undertaken under a decision-making framework called the "Unified Command System" (UCS). The UCS is an incident command partnership that ensures consultation and coordination among principal parties; in the case of marine casualties and potential jettison situations, the principal parties are the federal government acting through the on-scene coordinator; the state, acting through a predesignated representative; and the owner or other predesignated responsible individual. The UCS ensures, at a minimum, consultation among the parties before major decisions are made. It is the premise of the committee's recommendations that a decision to jettison would be made under the UCS.

SAFE HAVENS

A "safe haven," in the context of vessel salvage, is defined as a place of refuge where conditions are suitable for damage surveys and emergency repairs. Such refuge would entail shelter from winds, currents, and waves; adequate bottom conditions to provide a suitable anchorage; well-charted approaches; good beaching ground; and the potential for channel blockage. Hundreds of locations along the U.S. coast are acceptable safe havens, and one or more lie within reasonable reach of virtually any casualty. However, if a stricken vessel is in danger of spilling a hazardous cargo or fuel oil, if free flow of waterway traffic could be impaired, and if life-saving operations could be accomplished at sea, it appears unlikely that the Coast Guard or other federal, state, or local officials would permit the stricken vessel to be moved to a safe haven for repairs. This dilemma stems in part from the fact that conditions for suitable safe havens often also provide key habitats for biological populations, thereby increasing the risk of serious environmental damage should cargo be spilled. Only when a casualty occurs in an extremely sensitive area at the outset is movement to a safe haven likely to be approved.

The 1982 NRC report on marine salvage recognized the possibility that a salvor's request to move a ship to a safe haven might be problematic on environmental grounds. The 1982 committee pointed out that there are objective methods for evaluating relative environmental sensitivity and selecting among alternative sites to minimize environmental damage. That committee recommended that the Coast Guard develop criteria for safe havens and identify suitable sites, and that the regional response teams establish procedures to make safe havens available when needed. No significant action was taken on those recommendations, and the issue has become, if anything, even more contentious in the years since the 1982 report.

There have been only a few requests for access to safe havens in the past decade. Probably the most notable was the safe haven afforded the *Exxon Valdez* on its removal from Bligh Reef in April 1989. At the request of Exxon Corporation, the refloated vessel was moved so that temporary repairs could be undertaken before transit to the West Coast for reconstruction of the extensively damaged hull. In that instance, despite considerable cleaning of the ship's cargo holds, West Coast ports were reluctant to accept the vessel. Extensive negotiations were necessary before the ship was finally accepted in San Diego.

In October 1984, owners of the *Puerto Rican* were denied permission to bring the vessel into San Francisco Bay following an onboard explosion because oil was leaking from the vessel. While the ship was being towed out to sea, the stern section of the vessel parted from the bow and sank in 2,400 feet of water, resulting in chronic pollution for several months. When the bow section was determined to be structurally sound and free of further leakage, it was towed into San Francisco Bay, where 60,000 barrels of oil were offloaded.

The safe havens issue is not unique to the United States. It was addressed in the International Salvage Convention of 1989:

> Article 11. Cooperation. A State Party shall, whenever regulating or deciding upon matters relating to salvage operations such as admittance to ports of vessels in distress or the provision of facilities to salvors, take into account the need for cooperation between salvors, other interested parties and public authorities in order to ensure the efficient and successful performance of salvage operations for the purpose of saving life or property as well as preventing damage to the environment in general.

Few Coast Guard districts have formal policies regarding the approval of safe havens, and it appears that in almost all cases the decision to relocate a disabled ship that poses a potential hazard to the environment will be *ad hoc* and based on a complex set of factors. Interestingly, the one exception may be in Prince William Sound, where the policy is for the captain of the port (COTP) to designate safe havens as deemed necessary to minimize pollution effects and human hazards. While some coordination with other agencies is necessary, prompt and predictable action can be expected in Prince William Sound, Alaska, due to predesignation of safe havens. Unfortunately, denials of safe haven can be expected along most of the U.S. coastline. The lack of action on the part of the Coast Guard since the 1982 NRC report leaves the industry with few contingency planning options. The Coast Guard continues to rely on an after-the-fact analysis and determination by the COTP before considering the offering of safe haven. Moreover, if a vessel is losing oil, the COTP is likely to deny entrance, even though refusal may lead to a larger environmental catastrophe. The issue facing the COTP is potential pollution of one area in order to protect another. Without preplanning and analyzing alternatives, no or few COTPs can make that trade-off decision. To safely address the safe haven issue and provide the COTP with the necessary decision tools and information for emergency situations, a formal analysis program, such as the one for making a decision to use oil dispersants, is needed, with predefined areas for safe havens such as the tiered dispersant areas.

Findings

Predesignated safe havens do not exist within the United States, with the exception of Prince William Sound. Decisions concerning whether to provide safe havens are made on an *ad hoc* basis by the COTP in each case. The process for providing safe havens is not widely understood. As a result, potential safe havens may be sought following casualties when, in fact, none can be offered; or, a safe haven will not be requested even though it should be.

The lack of safe havens impedes traditional salvage procedures as a vessel probably cannot be considered "saved" until it is safely in port and secured. The absence of predesignated safe havens increases the risk of catastrophic outcomes having environmental consequences from marine vessel casualties.

SALVAGE READINESS OF VESSEL AND CREW

The concept of "salvage friendly" vessel design and training to ensure that a vessel and crew are well prepared and equipped to handle emergency situations has been adopted by some segments of the maritime community over the past 12 years. This approach, which initially focused on the facilitation of rescue towing, is expanding to embrace other aspects of vessel readiness, such as transfer piping between cargo tanks and empty spaces in a tanker. A salvage friendly approach to vessel design and crew training can have a significant impact on the outcome of salvage operations. In fact, it can easily make the difference between success and failure.

In July 1981, the Oil Companies International Marine Forum (OCIMF) published *Recommendations on Equipment for the Towing of Disabled Tankers.* In November 1983, the International Maritime Organization (IMO) adopted *Resolution A.535(13), Recommendations on Emergency Towing Requirements for Tankers.* The purpose of the IMO recommendations is to facilitate salvage and emergency towing operations on new and existing tankers, primarily to reduce the risk of pollution. The recommendation recognizes the need to harmonize the system components: tug or towing vessel, towline, pennant, chafing chain, fairlead, and towing gear connection or strongpoint on the vessel to be towed. The system should facilitate the ease of connection and be capable of being connected and released onboard the towed vessel in the absence of main power. The system should be standardized at the point of connection of the towline to the chafing chain.

Since the IMO resolution was published, sentiment has grow in favor of requiring tankers to have not only the towing component, but also means of rapidly deploying the equipment by a minimum number of persons aboard a vessel that has lost power. The IMO Subcommittee on Safety of Navigation is revising the resolution.

Vessels transiting Prince William Sound have been required to carry an emergency towing package that can be readily deployed. The towing systems have been implemented in various ways, and the time required for deployment ranges from 15 minutes to over three hours if vessel power is lost, according to the Disabled Tanker Towing Study Group (1993). That group also concluded that designs for quick deployment (15 minutes or less, with or without power) are necessary if these systems are to be of significant help in an emergency. In addition, the systems should be deployable by as few as two crew members, as most of the crew will be concerned with other aspects of the casualty.

While the focus to date has been on vessels carrying oil as cargo, emergency towing arrangements should also be considered for all vessels above a certain size. This strategy would address environmental concerns related to hazardous materials and the substantial amounts of fuel carried aboard cargo ships, as well as crew and passenger safety considerations. Such a safety measure could be applied through the development and promulgation of an international standard under the auspices of IMO.

Although the advantages of built-in equipment may be most obvious in cases of rescue towing, vessel modifications will also simplify a salvor's task. For example, cargo tank vent piping could be modified to provide standardized, valved, flanged connections that could be used in an emergency to connect the salvor's hoses (Stiehl, 1993).

Although making a vessel salvage friendly through structural and mechanical modifications is extremely helpful, crew training must not be ignored. Familiarity with salvage processes and procedures (described, for example, in *Peril at Sea and Salvage, A Guide for Masters* [International Chamber of Shipping and OCIMF, 1992]) can significantly improve the effectiveness of initial actions in emergencies. At least one major West Coast tanker owner, in concert with its salvage contractors, has developed programs to train personnel to react proactively in emergencies in a manner consistent with the salvage contractors' practices (Stiehl, 1993). The salvage contractors provided training and resources and specifications for structural modifications to vessels to make them compatible with the salvors' equipment.

Despite such positive steps, numerous vessels are still not in compliance with IMO Resolution A.535(13). The scarcity of salvage-friendly vessels, despite the recommendations made over 10 years ago, is due to the voluntary nature of compliance with IMO resolutions. A few regulations, such as those stemming from the TransAlaska Pipeline Act (P.L. 93-153), have required compliance, but the voluntary regime is prevalent internationally. Increases in liability damages may prove to be strong incentives for compliance in the near future.

Findings

The development of salvage-friendly vessels and crews can be a positive element in preventing a casualty or mitigating its effects in an emergency situation.

There is considerable history of salvage-friendly modifications to vessels, notably tankers. These alterations have been made primarily to facilitate rescue towing, but efforts are being expanded to include alteration of shipboard piping systems for the internal transfer of liquid cargo from damaged tanks.

Compliance with recommendations concerning salvage-related modifications is largely voluntary, and unfortunately there are still many tank vessels that have not been modified. There are no national or international recommendations for vessels other than tankers. Consequently, dry cargo and passenger vessels are seldom if ever fitted with emergency towing equipment.

THE NAVY'S CONTRIBUTION TO THE NATIONAL SALVAGE CAPABILITY

In addition to providing support for the U.S. Navy, the Navy fleet salvage resources and the Naval Sea Systems Command are permitted to provide back-up salvage capability for the private sector should available commercial resources be insufficient to handle a casualty. During the past 10 years, there have been few, if any, marine casualties in U.S. waters where commercial salvage equipment and expertise were insufficient. It can be assumed, therefore, that calls upon the Navy assist in or take over the salvage of a commercial vessel will continue to be infrequent.

The 1992 International Salvage Industry Survey states that international salvage resources are in serious decline, leading to the disappearance of a number of long-established salvage companies. A primary cause of the decline is "fierce competition from nondedicated resources, aggravated by the search for low-cost dry and wet salvage services by shipowners, insurers, and P&I Clubs" (Tecnitas, 1992).

It appears that current salvage resources have been sufficient to handle casualties in the last decade or so. However, the disappearance of traditional U.S. salvors and some of their equipment, particularly the dedicated salvage tug, is also a consideration. If an accident occurs where salvage resources are unavailable or inadequate, and Navy assets are available, then the Navy could be called upon to participate in commercial salvage operations.

Dedicated salvage vessels are expensive to build and maintain on a standby basis. Cutbacks in military resources following the end of the Cold War are reducing the Navy fleet's salvage assets, and the Navy's salvage force of the future will be tailored to a different world situation. Consistent with the Secretary of the Navy's recent white paper, *...From the Sea*, the Naval force will focus on operations in littoral waters.[12] In such operations, particularly in the two-region conflict contingency projected by U.S. military planners, Naval salvage forces become even more important than they are in a blue water (open sea) environment. However, it is unlikely that the projected Naval fleet salvage force level of 14 ships (seven T-ATFs, four ARS 50s, and three ATSs)[13] will survive into the next century. A reduction of some four to six T-ATFs is possible.

The five remaining World War II vintage salvage tugs (ARS 38 class) will be decommissioned by the end of fiscal year 1994. These ships are expensive to maintain and operate and with 3,000 SHP have insufficient power to be of much use in the rescue of large, modern commercial ships. In military service, they are manned by 6 officers and 97 enlisted men. These vessels are not adaptable to commercial salvage service.

The T-ATF 166 class of fleet ocean tugs, now operated by the civilian crews of the Military Sealift Command, are the most suitable Navy vessels for supplementing commercial resources. They have 7,000 SHP and can be operated with commercial-size crews. Their utility as salvage ships in their present configuration is limited, as they lack installed salvage equipment and lifting capability, and their crews receive

[12] The Navy's focus is shifting from high seas "blue water" operations to "power projection ashore."

[13] The Navy classifies its ships according to their capabilities. The ARS is equipped for rescue salvage, the ATS for towing and salvage, and the T-ATF for towing and firefighting.

only limited salvage training. T-ATFs are designed, however, to accept portable salvage and diving systems as well as numerous specialized military and commercial oceanographic systems, and they can berth and feed 14 persons in addition to their normal crew.

Navy SUPSALV maintains an inventory of salvage equipment in its worldwide network of Emergency Ship Salvage Material (ESSM) facilities. This equipment is designed and palletized for transport to and loading aboard the T-ATFs. Fleet Mobile Diving and Salvage Units (MDSUs) maintain Fly-away Dive Systems (FADS) and salvage equipment ready for deployment aboard T-ATFs. The T-ATF is versatile, and when outfitted to meet the needs of a specific operation, it can fulfill the mission of several specialized ships in salvage, deep sea diving, deep ocean search, ROV operations, ocean towing, or firefighting. Presently, three T-ATFs are based in Norfolk, Virginia, two in San Diego, California, and one each in Guam and Singapore.

Two or three T-ATF 166-class ships, strategically located and properly outfitted and backed up by ESSM facilities, could restore much of the lost traditional salvage capability along the coasts of the continental United States.

One method by which these vessels could be made available for commercial work would be to provide them on a bareboat charter basis to commercial operators, who would keep them on standby to supplement existing commercial assets while also performing Navy work within a designated area. Alternatively, private salvage companies could maintain the vessels, in the manner of the two commercial vessels based on the West Coast, with a skeleton crew backed up by a full crew that can be assembled in two hours. In either case, the contract could be written to allow immediate redelivery to the Navy in the event of a national emergency with or without the civilian crew. This approach would require legislative and/or regulatory changes and the consent of both the Navy and existing commercial operators.

The Navy scaled back and ultimately abandoned similar commercial arrangements in the 1970s due to escalating costs and continued pressure on limited budgets. (A major cause of the problem then was the decline in commercial salvage business, which had been used to offset out-of-pocket costs to the Navy. That situation has not changed and is probably worse now from a salvage business viewpoint, although it is much better from the shipowner's and the environmental perspective.) For such arrangements to be viable now, nondiscretionary Congressional direction and a reliable, long-term source of funding would be required.

Absent such a commitment from the Congress and the administration, any improvement on the *status quo* is unlikely. The Navy salvage community will continue to compete with other important defense programs for funding. Unless the legislative and executive leadership take a strong stand on this issue, it is likely that Navy salvage assets will be reduced further. Moreover, without a national commitment, it is unlikely that the Navy will be provided the resources for increased salvage protection in areas beyond its own operating venues.

Findings

The Navy has significant floating, dedicated salvage assets that are not generally available to the private sector. Some of these assets are being phased out as the Navy alters its salvage mission following the end of the Cold War.

Surplus assets, particularly the T-ATF class of ships, if operated by the private sector and strategically deployed, could go a long way to restoring the traditional salvage capability of the United States, particularly in rescue towing. The operation of these vessels by the private sector would require substantial subsidy, as it has been demonstrated in the United States and elsewhere that salvage revenues cannot cover the costs of operating and maintaining the vessels and their crews. The excess costs could be covered, as they were in the past, through the Salvage Facilities Act, and the plan could be implemented through the arrangements in place for Navy contracting for commercial salvage services. Nondiscretionary Congressional direction and a long-term source of funding would have to be obtained before this approach could be considered.

4

CONCLUSIONS AND RECOMMENDATIONS

Effective, time-critical salvage provides an important safeguard against environmental damage and commercial loss. It can prevent breakdowns, navigation errors, and other vessel mishaps from developing into more serious incidents, which could result in major pollution or channel and port closures. The salvage industry continues to face immense economic pressures. In general, the incidence of maritime casualties, and thus the need for salvage services, has declined by one-third since 1973. These trends are favorable for safety and environmental protection, but they have a negative effect on business conditions in the marine salvage industry.

Additionally, the nature of salvage has changed in the last decade. At least four major trends have been observed:

1. The motivation for maintaining a salvage capability has shifted from a private concern protecting the vessel and its cargo to a more public or societal interest in protecting the environment and economy from impacts of a vessel casualty.
2. The traditional dedicated salvage company, with fully integrated capabilities, has had to supplement its income by utilizing its salvage assets and experienced personnel in more conventional marine business. Salvage has become only one segment of the organization's business.
3. Compensation for salvage has not kept pace with either the changing roles of the salvor or the increased financial risks.
4. The salvor no longer is in charge of the decision-making process employed in responding to a marine casualty, particularly one involving pollution or the threat of pollution. Instead, the salvor's role is to assist and provide direction to a Unified Command System (UCS) involving the federal government, state government(s), and the responsible party (vessel or cargo owner or designee).

As a result of these trends, the U.S. salvage capability today is found in a small number of professional salvage companies, some dealing only with a specialized and limited aspect of salvage. Nevertheless, response to vessel casualties in the United States since 1982 has been timely and effective. The committee did not identify a pattern of failure to respond because of the lack of capability.

NATIONAL SALVAGE POSTURE

Despite the good response record of salvors in recent years, the committee perceives possible deficiencies in U.S. salvage capabilities. Conclusions regarding these potential deficiencies, and recommendations for addressing them, are outlined in the following sections.

COMPENSATION FOR SALVAGE

The committee concludes that, because of the change in societal values as they relate to salvage from an emphasis on preventing or mitigating the loss of vessels and cargo to concern for preventing or mitigating environmental damage and other indirect consequences the economic basis for rendering salvage services needs to be restructured. The committee therefore recommends:

The criteria for determining compensation for salvage in the United States should be updated to reflect changes in the business structure of the salvage industry and in societal values with respect to salvage and to incorporate the criteria delineated in the Salvage Convention of 1989.

The Salvage Committee of the Maritime Law Association should promulgate an updated regime of criteria for salvage awards, reflecting current salvage conditions, to be brought to the attention of courts and arbitrators without waiting for a case to be litigated or arbitrated.

NATIONAL SALVAGE POLICY

The national salvage policy continues to be that stated in the Salvage Facilities Act of 1948: that the public interest is served by maintaining salvage capability to provide for the national defense. In contrast to the 50-year-old statement of national policy, the committee concludes that salvage fulfills additional functions--preventing or minimizing marine pollution, providing for public safety, and minimizing the disruption of port activities. The committee therefore recommends:

The Congress should update the national salvage policy to ensure that an adequate level of salvage capability is present in U.S. waters. The policy should clearly delineate the following goals: to protect national security, to minimize or prevent environmental impacts due to pollution from marine casualties, to protect public safety, and to ensure minimal economic disruption resulting from marine casualties in the nation's ports and waterways.

The updated national salvage policy should specifically address the role of government agencies and their relationship to the private sector. At the very least, the policy should codify or clarify the current roles of the U.S. Coast Guard and the U.S. Navy, as follows:

- **The U.S. Coast Guard should oversee or direct response to marine casualties in which there is peril to life, the environment, or other public interests.**
- **The U.S. Navy should advance salvage technology and practice through technical development programs; provide technical assistance in nongovernmental marine casualties requiring salvage; maintain contractual arrangements with private contractors for salvage services in excess of the Navy's internal capability and make such services available on request; and train U.S. Navy and other personnel in salvage.**

The above functions are essential to government roles to ensure that adequate salvage capability exists in the United States and that there is adequate salvage response to marine casualties. Implementing this recommendation would, in the main, provide statutory authority for current practice.

The national salvage policy should clearly state that government assets will be used only when commercial assets are not available.

The 1982 National Research Council study on salvage recommended that the salvage issue be revisited in 10 years. This study coincides with that time frame and finds significant changes requiring action that justify the 1982 recommendation. The committee therefore recommends that the national salvage posture be assessed again ten years hence.

SALVAGE ASSETS AND SERVICES

The committee deems it to be in the national interest to ensure that the U.S. salvage capability remain (at a minimum) at its current level. However, the committee's analysis of marine risks, casualty rates, and salvage business conditions supports the conclusion that the current level of capability cannot be sustained, nor identified gaps filled, without attention to acquiring, maintaining, sustaining, and positioning salvage assets, and training and retaining salvage personnel. Furthermore, the committee concludes that these measures require the direct assistance of both the marine transportation industry and government. The committee therefore recommends:

The U.S. Coast Guard, in consultation with the U.S. Navy Supervisor of Salvage, should develop and promulgate a list of active salvors in accordance with the general criteria proposed by the committee. The list could be used to prequalify salvage companies for inclusion in vessel owners' contingency plans for casualty response and may include salvors that specialize in specific aspects of salvage. As a matter of corporate policy, companies needing salvage services should endeavor to contract for salvage services with professional salvage companies.

As new offshore tugs and escort vessels are built, designers and owners should be encouraged to add features to expand the response capability of those vessels to include some aspects of salvage. At a minimum, firefighting capability should be installed, as well as basic salvage equipment lockers.

Owners of all commercial vessels should be required to demonstrate that they have considered and are prepared to address, by contract or other approved means, key aspects of salvage capability as it relates to the safe conduct of their vessels in U.S. waters. (At present, this requirement applies only to vessels carrying petroleum as cargo.)

TUG AVAILABILITY

The committee concludes that geographic gaps may exist in the availability of adequate towing vessels for control and management of stricken ships, particularly in the Florida Straits and in areas of the West Coast. In light of physical and economic limitations to providing dedicated rescue towing salvage vessels, the committee recommends:

The U.S. Coast Guard should expeditiously complete studies mandated by OPA 90 on vessel-routing schemes and exclusionary zones and vigorously pursue instituting, as appropriate, the conclusions arising from those studies in both national and international forums.

SALVAGE READINESS OF VESSEL AND CREW

The voluntary nature of compliance with "salvage-friendly" vessel designs, such as that proposed by IMO Resolution A.535(13), has not had the expected result of widespread adoption of such features and the potential increase in the ability to save vessels in distress. The committee therefore recommends:

The Coast Guard should urge the International Maritime Organization to proceed with the revision of Resolution A.535(13) "Emergency Towing Requirement for Tankers" as expeditiously as possible and to promulgate the revised resolution in the form of a regulation requiring compliance by new and existing vessels. The Coast Guard should request that the IMO consider requiring emergency towing arrangements on vessels other than tankers.

In addition, the Coast Guard should urge vessel owners to incorporate other "salvage-friendly" modifications, such as cargo piping system modifications, and institute formal salvage training for vessel crews.

TRAINING

The committee concludes that, with few opportunities to practice the salvage profession, the next generation of salvors is not being trained or attracted to the industry. The committee therefore recommends:

The U.S. Navy training facilities at Panama City, Florida, should be made available to selected industry and/or federal and state agency personnel sponsored by professional salvage companies and the respective agencies. Such training could be offered on a total reimbursement basis or financed by the Salvage Facilities Act.

The salvage and marine transportation industries should establish salvage training at a maritime institution on a cooperative basis (as was done with maritime firefighting schools). In addition, they should cooperatively develop a career track program within and across their respective industries to attract individuals to the salvage industry.

MARINE FIREFIGHTING

The committee concludes that marine firefighting experience in port areas is lacking. There are significant differences in marine firefighting capabilities among port areas. Of particular concern is the interface between local fire departments and the marine transportation system. The committee further concludes that the location of prepositioned marine firefighting systems relative to vessel traffic patterns is a concern in some geographic areas. Of special concern are ports and areas with high passenger vessel concentrations. The committee therefore recommends:

The area planning process within the National Contingency Plan should include a review of local and area firefighting and salvage readiness and capabilities.

RESPONSE TO HAZARDOUS CARGO

The committee concludes that capability for responding to casualties involving hazardous cargo is limited. The committee therefore recommends:

The Coast Guard or another government agency should undertake a study on response to marine casualties involving hazardous cargo and the role of the salvor in the response.

JETTISONING

The committee concludes that ambiguities in federal and state oil pollution laws have created uncertainty concerning liability for acts of jettison. The committee therefore recommends:

The National Contingency Plan should be modified to incorporate criteria that would authorize intentional jettison when necessary. Modifications should also clarify the Federal On-scene Coordinator's ability to authorize incidental discharges resulting from ongoing cleanup operations, such as decanting from skimmers and pumping of engine rooms, through the daily work plans approved by the Unified Command System where appropriate.[1]

SAFE HAVENS

The committee concludes that, in the absence of predesignated safe havens, the risk is increased of marine casualties having catastrophic outcomes with environmental consequences. The committee therefore recommends:

The Coast Guard should promulgate the process by which a "safe haven" is identified. To the extent possible, area plans should evaluate candidate sites for potential safe haven areas.

[1] The jettison issue was addressed by the committee in a separate report (NRC, 1994), which appears, with more specific recommendations, in Appendix B.

APPENDIX A

BIOGRAPHIES OF COMMITTEE MEMBERS

GORDON W. PAULSEN, the committee chair, practices admiralty law in New York City with the firm of Healy & Baillie. He is a member of the Permanent Advisory Board of the Admiralty Law Institute (Tulane University), the Maritime Law Association of the United States, and the Comité Maritime International. He has served as chairman of a number of distinguished maritime safety panels, including what is now the Navigation Advisory Council of the United States Coast Guard; the Rules of the Road Advisory Council and its successor, the Navigation Safety Advisory Council; and the Committee on Navigation and Coast Guard Matters of the Maritime Law Association. A recipient of the Meritorious Public Service Award and the Distinguished Public Service Award, Mr. Paulsen has been recognized for his contributions to navigation safety. He also provided testimony to the Alaska State Oil Spill Commission regarding the *Exxon Valdez* spill. Mr. Paulsen was a member of the NRC Committee on the National Salvage Posture. He received his BS and MA from New York University and his LLB from Columbia University. He is a member of the New York and American bar associations and the American Arbitration Association, and he is a fellow of the American College of Trial Lawyers.

PETER BONTADELLI is administrator of the California Department of Fish and Game's Office of Oil Spill Prevention and Response. He has primary authority to direct prevention, removal, abatement, response, containment, and cleanup efforts related to oil spills in California's marine waters. His prior experience at the Department of Fish and Game included service as special assistant to the director, chief deputy director, and, most recently, department director, a post he held for five years. During that time he served on various distinguished environmental panels, including the Pacific Flyway Council (he is a past president), the North American Wetlands Conservation Council (he was appointed by the U.S. Secretary of the Interior), the Pacific Fishery Management Council, the Pacific States Marine Fisheries Commission, the International Association of Fish and Wildlife Agencies, and the Western Association of Fish and Wildlife Agencies. Mr. Bontadelli received his BA (political science) from the University of California at Davis in 1970.

J. HUNTLY BOYD, JR. is a senior associate at Booz-Allen & Hamilton, Inc., an international engineering firm. Prior to joining Booz-Allen, he was a private consultant conducting marine engineering studies and casualty analyses, and he completed 28 years of service in the U.S. Navy, finishing his Naval career at the rank of captain. While in the Navy, he served as director of ocean engineering and supervisor of salvage in the Naval Sea Systems Command and commanded a naval shipyard. Mr. Boyd is an honorary life member of the American Society of Naval Engineers and an active member of the Society of Naval Architects and Marine Engineers (SNAME) and the Marine Technology Society (MTS). He is chairman of both the SNAME Panel on

Salvage and Rescue Systems and the MTS Marine Salvage and Towing Committee. Mr. Boyd received his BS from the U.S. Naval Academy and his MS (naval architecture and marine engineering) and professional degree (naval engineer) from the Massachusetts Institute of Technology.

KENNETH J. FULLWOOD is a master mariner (British Foreign Going Certificate) with 40 years experience in the tanker industry, 18 at sea on tankers, including 3 as master and 7 as chief officer on various vessels, including crude oil tankers over 100,000 deadweight tons. He is manager of Maritime Relations, Environmental Affairs, Safety & Nautical Services for Mobil Shipping and Transportation Co., with responsibility for the company's worldwide marine operational safety and environmental protection policies. His department provides technical advice regarding worldwide tanker and terminal operations, monitors government initiatives affecting marine activities, and develops appropriate industry response in concert with such organizations as the Oil Companies International Marine Forum, the American Institute of Merchant Shipping, and the American Petroleum Institute. At Mobil, he has studied tank cleaning problems in Very Large Crude Carriers, an effort that formed the basis of Mobil's tank cleaning and antipollution policies. He also was in charge of successful salvage and pollution response following the stranding of the 30,000-ton tanker *Mobil Oil* in the Columbia River in 1984 and the refloating of a sunken tugboat in New York Harbor. Born and educated in England, Captain Fullwood has been a U.S. citizen for 10 years.

RICHARD LEE is professor of oceanography at Skidaway Institute of Oceanography in Savannah, Georgia. His academic career of 30 years has included teaching and research posts at Scripps Institution of Oceanography, California State University at San Diego (UCSD), Pennsylvania State University, and Georgia Institute of Technology. His research has focused on the biochemistry of marine organisms and the fates and effects of chemical compounds on marine biota, especially bioaccumulation and metabolism of hydrocarbons. He served as chairman of a workshop on the problems of monitoring biological effects of pollution at sea for the International Council for the Exploration of the Sea, and as chairman of a panel at the Gulf-South Atlantic Region Ocean Pollution Conference sponsored by the National Oceanic and Atmospheric Administration. He contributed to *Petroleum in the Marine Environment*, a report by the National Academy of Sciences, and he served on the Science Advisory Board, Ecological Monitoring Subcommittee, of the Environmental Protection Agency. He has published more than 100 articles in professional journals. Dr. Lee received his BA and MS (chemistry) from the California State University at San Diego (UCSD) and his Ph.D. (marine biology) from UCSD, Scripps Institution of Oceanography.

J. H. "MICK" LEITZ has been in the marine and heavy marine salvage business since 1959. He is president and part-owner of Fred Devine Diving and Salvage Co., which operates the dedicated salvage vessel, the *Salvage Chief*; president of Leitz Marine Recovery Systems, a company involved in the design and construction of deep-water recovery systems; and is president and senior salvage master for J.H. Leitz & Associates, Inc., a salvage and marine consulting firm. He has been responsible for conceptual planning and engineering and personally supervised many salvage operations. As a salvor, he has responded to ship fires, explosions, and other casualties on many vessel types, including jack-up drilling rigs, tankers, and ocean barges from the Arctic Ocean to the Gulf of Mexico and the South Pacific. His accomplishments include successful salvage of the *Exxon Valdez*, an operation that involved safely pumping one million barrels of oil off the grounded ship, injecting inert gas into the ruptured tanks, and refloating, repairing, and safely delivering the vessel without further pollution to drydock in San Diego.

JOHN H. ROBINSON is director, Gulf Program Office of the Office of the Chief Scientist, National Oceanic and Atmospheric Administration (NOAA). In this position he directs NOAA research activities associated with the aftermath of the Persian Gulf war marine science programs to assess the impact of the oil spills in the Persian Gulf and research programs to investigate the impact of oil fires in Kuwait. Previously, he was manager of NOAA's Hazardous Materials (HAZMAT) Response Division, where he developed and managed the agency's spill response and hazardous waste site research program; established regional scientific support programs in U.S. coastal areas; and served as scientific coordinator for the Ixtoc I, *Exxon Valdez*, and numerous other oil and chemical spills. While at HAZMAT, he originated a program for Computer-Aided Management of Emergency Operations. He began his government career in 1964, working for the Apollo manned-spacecraft program. Robinson received his BS degree (industrial engineering) from Texas Tech University in 1961.

NINA SANKOVITCH was until recently a senior project attorney with the Natural Resources Defense Council. She worked with the council's Coastal Project since 1989, focusing on oil spills, ocean dumping, and coastal issues. She is a member of New York Governor Mario Cuomo's Task Force on Petroleum Spill Emergency Protection, Subcommittee on Oil Spill Response. She served on a number of other committees related to waste management in the environment, including the Safe Harbor Coalition, dedicated to the protection and enhancement of New York Harbor (serving as chairperson); the Steering Committee of the Clean Sludge Coalition, devoted to the environmentally sound management of sludge on land; and the Citizens Advisory and Technical Advisory committees to New York City on land-based sludge management. Ms. Sankovitch received her BA from Tufts University and JD from Harvard Law School.

ROGER VAN DUZER is manager of Marine Operations for Shell Oil Co.'s Marine Department in Houston, Texas. This position involves ensuring safe, efficient, and environmentally responsible marine transportation for Shell, which is a major charterer of tankers and barges. He has held various positions with Shell in the Marine Technical Division, Port Operations, and other business management positions, in addition to spending two and a half years with Shell International Marine in London. Prior to joining Shell in 1976, he was an officer in the U.S. Navy. Mr. Van Duzer's Naval career included sea duty on a destroyer (as a deck officer) and on an aircraft carrier (as an engineer). His shore duty included service as the diving and salvage officer at the Boston Naval Shipyard, a position that provided extensive experience with new construction and repair of Navy ships. He was qualified as a Navy Engineering Duty Salvage Officer in 1969. Mr. Van Duzer is a graduate of the U.S. Naval Academy and received his MS (naval architecture) and professional degree (naval engineer) from the Massachusetts Institute of Technology.

JOHN A. WITTE is president and chief executive officer of Donjon Marine Co. of Hillside, New Jersey. Mr. Witte practiced admiralty law for four years prior to managing and expanding Donjon Marine, a marine salvage and marine equipment sales company he incorporated in 1966. His extensive experience in the marine salvage industry originated in his father's vessel demolition, marine salvage, and wreck removal business. His experience running Donjon Marine has included service as salvage master and hands-on work in all aspects of salvage operations, including rigging, welding, diving, underwater cutting and burning, and controlled explosives.

He has worked on salvage operations all along the U.S. East Coast, the Gulf of Mexico, the Pacific Ocean, and the North Sea, in varied and hazardous environments from shallow water to the deep ocean, in strong currents, and in crowded harbors. He has overseen diverse projects including the raising and removal of tugboats, a sunken Liberty ship, and tankers, efforts that required pollution abatement, cargo transfer, and detailed pumping schemes. Mr. Witte received his BS (management) from Lehigh University and his JD from Villanova University School of Law.

APPENDIX B

JETTISON REPORT AND RECOMMENDATIONS

The Symposium on the Purposeful Jettison of Petroleum Cargo was convened on February 23, 1993, by the Committee on Marine Salvage Issues of the National Research Council, as the first phase of the committee's assessment of marine salvage in the United States. Participants represented federal and state agencies and the fields of marine environmental protection, salvage, vessel operations, admiralty law, and marine insurance. Their purposes were to:

- Assess the significance of the jettison issue and its implications for shipping and marine environmental protection.
- Document the need to clarify U.S. law concerning intentional discharges of petroleum cargoes to save ships and prevent the loss of larger amounts of cargo.
- Consider the implications of advances in oil spill contingency planning, environmental data acquisition, and spill trajectory forecasting, especially how such advances might be harnessed in making time-critical operational decisions about stranded tankers.
- Make recommendations concerning the feasibility of developing guidelines for deciding whether to discharge oil intentionally, including consideration of other options.

This report has three parts. The first is a summary of the symposium proceedings, including presented papers, audience question-and-answer sessions with the speakers, and two panel discussions that focused on jettisoning as an option for response to an accident scenario. The second section of the report outlines the committee's analysis of the major issues emerging from the symposium. The third section presents the conclusions and recommendations the committee derived from its analysis. In developing its analysis, conclusions, and recommendations, the committee received some input from sources other than the papers and discussion in symposium.

SUMMARY OF SYMPOSIUM PROCEEDINGS

History and Technical Background

Michael Ellis, general manager of the Salvage Association, noted that the practice of jettisoning cargo to lighten a ship dates back to Biblical times. In 900 B.C., the laws of the island of Rhodes prescribed that cargo jettisoned for reasons of common safety would be made good in General Average (i.e., the interests benefited by the jettison would be obliged to share the loss), a rule that persists today. Jettison has continued to play a role in salvage in the twentieth century, but the cargo involved usually has been dry bulk. No classic cases of jettison of part of an oil cargo, with successful salvage of the remainder, have been identified in the last thirty years or so, although tankers often are lightened for refloating.

The only purposeful jettison of oil cargo identified in recent years was from the *Zoe Colocotronis*. After grounding off Puerto Rico in 1973, some 5,000 tons of crude oil were jettisoned and the ship was refloated. The case is notorious, Ellis said, because such drastic action now appears to have been unnecessary. When the *Arrow* grounded off Nova Scotia in 1970, jettisoning was authorized, but part of the vessel sank before action could be taken. After the *Argo Merchant* ran aground on Nantucket Shoals in 1976, jettisoning was suggested but rejected. The vessel eventually was lost.

According to Ellis, salvors are reluctant to jettison even when it may be more expedient than waiting for proper equipment and weather to lighten the ship. He cited three reasons:

1. Salvors are concerned about the environment and want official approval to jettison.
2. Discharge of valuable oil cargo means a reduction in the salved values and perhaps in the salvage award.
3. Circumstances of groundings rarely support a clear-cut decision to jettison.

Nonetheless, in cases where tugs cannot refloat a tanker and lightering is not possible, it may be "better to jettison and accept that sometimes sacrifice is necessary for the common good," Ellis said.

F.R. Engelhardt, vice president for research and development for the Marine Spill Response Corporation, discussed environmental risk as a function of oil spill size. Variables that can influence risk include characteristics of the oil, physical environmental conditions, containment and recovery measures, geology of the impact zones, toxicological sensitivity of vulnerable species, and ecological characteristics of vulnerable areas. These variables interact to drive weathering rates and persistence, spread of the spill, direction of slick movement, effectiveness of response measures, size of the impact zone, extent and duration of biological effect, and degree and rate of recovery. Thus, large spills do not necessarily have greater potential for environmental impact than do smaller spills. Engelhardt said the spill size influence should be analyzed in depth based on the global record for marine spills, perhaps using a proposed marine oil spill scale similar to the Richter scale for earthquakes.

Jerry Galt, a physical oceanographer at the National Oceanic and Atmospheric Administration (NOAA), described oil spill trajectory modeling. He outlined how these models work, the types of models available, factors that limit model accuracy, how models are used, the technology available to support models, and needed improvements. Models can be used to estimate some aspects of currents and the projected distribution of jettisoned oil. Models are not a tool for arriving at a definitive recommendation to jettison, Galt said, but they can establish that a situation will deteriorate if such action is not taken. "It is an exploration of the situation – to find out the worst downside and then plan for that scenario," Galt said.

Legal Problems Concerning Jettison

Warren L. Dean and Laurie L. Crick, a senior partner and an associate, respectively, in the law firm of Dyer, Ellis, Joseph & Mills, summarized laws affecting the jettisoning of oil. Under current laws, a salvor does not qualify for any award unless some property is saved – thus the "no cure, no pay" principle traditionally embodied in "open form" salvage agreements. Furthermore, the award may be reduced to compensate the vessel owner for losses caused by the salvor's failure to exercise due care. The degree of care required varies with the source of the risk. The salvor is liable for losses due to the salvor's gross negligence or willful misconduct; the salvor also is liable for "distinguishable and separate"[1] damage to property caused by failure to exercise ordinary care.

The United States is a party to the International Convention for the Prevention of Pollution from Ships, 1973, and its 1978 Protocol (MARPOL), which generally prohibit pollution of the oceans from seagoing vessels but make exceptions for salvage-related discharges. Discharges that are necessary to secure the safety of the ship or to save lives at sea, or that result from damage to a ship or its equipment are exempted from the treaty provisions.[2] Until recently, this exception immunized salvors from liability for discharges at sea, but not for discharges within the three-mile U.S. territorial sea, where the purposeful jettisoning of oil is prohibited under the Federal Water Pollution Control Act (FWPCA). The FWPCA prohibits discharges of oil in quantities sufficient to cause a sheen on the water and imposes civil and criminal penalties for unauthorized discharges, including those into waters under state jurisdiction. However, the FWPCA does not prohibit discharges into the Exclusive Economic Zone (EEZ)[3] if they are permitted by MARPOL.

The Oil Pollution Act of 1990 (OPA 90) conflicts with MARPOL protection for discharges within the EEZ because OPA 90 makes the vessel owner and operator responsible for removal costs and damages for all discharges into U.S. navigable waters and the EEZ. Consequently, the owner's and operator's liability must be taken into account when conducting salvage activities. Outside the territorial sea and within the EEZ, OPA 90 does not actually prohibit salvage-related discharges, so the salvor, owner, and operator of a vessel are not subject to civil or criminal penalties. Under OPA 90, however, liability for damages and removal costs would be incurred even if the discharge were permitted by MARPOL.

Anyone who negligently or knowingly discharges oil in violation of the FWPCA is subject to criminal penalties. Therefore, a salvor may be subject to prosecution if cargo is jettisoned, even if this action prevents loss of the entire cargo and additional environmental damage. As of February 1993, there had been no legal cases involving a salvor who jettisoned oil into the navigable waters of the United States.

OPA 90 provides limited immunity ("responder immunity") to persons "rendering care, assistance, or advice consistent with the National Contingency Plan [NCP] or as otherwise directed by the President." It is unclear whether a salvage-related discharge can be consistent with the NCP, which is supposed to "minimize damage from oil and hazardous substances discharge." A salvor should be eligible for conditional immunity under this provision, Mr. Dean and Ms. Crick said. To ensure responder immunity, the salvor may need to obtain permission to discharge from the President (whose authority devolves to the federal on-scene coordinator [FOSC]). But obtaining such permission may be difficult and impractical in a salvage situation. As

[1] This term refers to damage distinct from that caused by the original peril.

[2] Citations for these exemptions and for the various laws mentioned in the summary report may be found in the authored papers.

[3] The EEZ, created by a 1983 presidential proclamation, is a belt of jurisdiction over seabed resources adjacent to the United States and its island territories, extending 200 nautical miles from shore. The proclamation extends U.S. sovereign rights in this region for exploration, utilization, conservation, and management of natural resources.

of February 1993, there had been no legal cases where a salvor or vessel owner had been held directly responsible for a jettison directed by a salvor. In the aforementioned case of the *Zoe Colocotronis*, the vessel owner was fined $5,000 under the FWPCA and was held responsible for the resulting environmental damage and removal costs. (The master had made a unilateral decision to jettison.)

OPA 90 preserves the right of states to impose liabilities and obligations above federal requirements, and most coastal states have enacted oil pollution liability statutes. In general, these laws impose liability – often unlimited – on anyone who discharges oil. Thus, discharges into U.S. navigable waters may violate state law.

The International Convention on Salvage 1989, which will take effect when ratified by fifteen nations (six had done so as of the time of the symposium), emphasizes the salvor's duty to protect the environment and authorizes special compensation to promote that duty. This special award may be added to conventional awards that are insufficient to cover expenses, or it may be used to reward actions that prevent or minimize damage to the environment.

In conclusion, Mr. Dean and Ms. Crick said that because salvage was not discussed in the legislative history of OPA 90, the Congress apparently did not consider fully the implications for salvage, and that full assessment of these implications awaits judicial and regulatory interpretations of the Act. "What is clear, however, is that salvors must now consider carefully their environmental responsibilities . . . [and] salvors and the rest of the maritime community [will have] to monitor the implementation of these new laws so that the new environmental obligations of salvors do not undermine the certainty of the legal regime upon which salvage operations necessarily rely."

Philip A. Berns, U.S. attorney in charge of the West Coast office of the Torts Branch, Civil Division, commented on various aspects of liability. He agreed with Dean and Crick that the courts must resolve many of the relevant issues. He said the decision of the FOSC is a significant factor in whether a salvor is liable for jettisoning.

Frederick F. Burgess, Jr., an attorney specializing in maritime and environmental law with Leboeuf, Lamb, Leiby and MacRae, reviewed the authority of the FOSC and the requirements of the NCP as well as a salvor's standard of care under various circumstances. For a jettison "directed" by the FOSC (as distinct from an unlawful discharge) under OPA 90, neither the salvor nor the federal government is liable, either for penalties or for removal costs and damages. Furthermore, jettisons consistent with the NCP invoke responder immunity and possibly the same immunity as "directed jettisons," at least under federal law.

Jettison of cargo to prevent an even greater spill is not authorized explicitly by either OPA 90 or the NCP, nor has the U.S. Coast Guard given any additional significant direction on this matter through commandant's directives or other instructions. However, under OPA 90 the President has the authority to destroy a vessel discharging or threatening to discharge oil; such destruction, of course, probably would result in discharge of oil into the sea, yet it is not an unlawful discharge. Therefore, jettisoning cargo to prevent the loss of a ship or to prevent the loss of a much larger amount of cargo also must be authorized. Consequently, if "directed jettisoning" is part of "removal" to prevent a substantial threat, it is authorized under OPA 90. Burgess also emphasized that, under OPA 90, actions to minimize damage from oil need be consistent with the NCP only "to the greatest extent possible."

OPA 90 requires the President to revise the NCP to meet the new objectives of the law. The revised NCP must include criteria and procedures for identifying and responding to a substantial threat of discharge, as well as procedures and standards for mitigating, preventing, or cleaning up an actual discharge. Mr. Burgess suggested that if the jettison tool is to be made available, the NCP be revised to "ensure that criteria, standards, and procedures are in place and exercised by the National Response Team, regional response teams, [the commandant], and [FOSC] to consider the jettisoning possibility expeditiously and make a prompt decision to direct this action if, in the judgment of the President's delegate, it will mitigate or prevent substantial threats of discharge." According to Burgess, "Government paralysis can cause far more serious consequences than a wise jettison decision."

State laws probably cannot impose criminal or civil sanctions on a salvor who jettisons oil at the direction of the FOSC because of the federal supremacy doctrine established by the courts, Burgess said. According to this doctrine, a state statute is void when it impedes the accomplishment and execution of the full purposes and objectives of the Congress.

Under the current NCP, a salvor who independently jettisons oil runs the risk of authorities disagreeing with that action, and the salvor therefore should seek direction or explicit agreement from the FOSC, Burgess said. If the revised NCP set forth circumstances when a vessel could be destroyed or cargo jettisoned, such actions more clearly could be considered "directed" by the President and would invoke full protection from liability for the salvor. Actions merely "consistent" with the NCP may not be as well protected. Mr. Burgess also proposed that The Commandant Instruction regarding intervention be revised to address more explicitly the jettison option; that the "harmful quantity" definition in the FWPCA be amended to permit jettisoning or other pumping of small quantities of oil under carefully prescribed circumstances, to achieve OPA 90 mitigation and prevention goals; and that a review be conducted of past casualties where the Coast Guard has intervened, to determine circumstances when jettisoning has or might have been used, and the consequences.

The standard of care for salvors depends on the type of injury, the injured party, and the applicable law. Under current U.S. law, only gross negligence or willful misconduct will make a salvor directly liable to third parties for removal costs and damages, according to Burgess. The most recent notable demonstration of this principle was the case of the *Amoco Cadiz*, where claimants damaged by discharged oil were denied any recovery against the salvor because no gross negligence or willful misconduct was found.

Robert H. Nicholas, Jr., general counsel of Exxon Shipping Co., addressed the need for and reasoning behind the 1989 Salvage Convention. Under current U.S. law, a salvor does not receive an award for preserving the property of third parties (e.g., the environment). Instead, the courts base awards on the degree of danger, the value of property saved, the risk incurred by the salvor, the skill and energy displayed, the value of the salvor's property, and the time and labor expended. If the focus of a salvor's efforts is to be redirected toward pollution prevention and mitigation, then the law must be changed to provide for adequate compensation to salvors who assume such risks, Mr. Nicholas said. The U.S. ratification of the 1989 Salvage Convention appears to be a step in this direction.

The 1989 convention also incorporates the concept of "liability salvage," under which the salvor may recover expenses from the tanker owner if the salvor prevents or mitigates pollution, even if the vessel is lost or only partially saved. Even though the

1989 convention is not yet in force, liability salvage has gained limited acceptance in the maritime community, in that it has been incorporated into open form contracts and marine insurance policies, Mr. Nicholas said.

V. Lee Okarma Rees, an attorney with Graham & Dunn, commented on the relationship between federal and state oil pollution laws and legal issues regarding the jettison of cargo. State laws may be preempted if they conflict with federal law, but Ms. Rees found no actual conflict with regard to jettison. States apparently may impose liability on the salvor for jettisoning and are not preempted by federal law, although there appear to be limits on that authority. Most states provide limited immunity for response contractors, but usually certain conditions must be met.

If a salvor jettisoning oil thereby prevents an even greater oil spill, then that salvor arguably should be entitled to an award in proportion to the potential liability avoided by the shipowner, Ms. Rees said. Given the new emphasis on avoidance of environmental damage in the 1989 convention, the courts could acknowledge liability salvage as a legitimate part of a salvage award. In conclusion, Ms Rees noted that "Shipowners and public and private salvors face uncertainty regarding potential liability for jettisoning cargo, even if the salvor's actions may be in the public interest by avoiding greater discharge of oil and greater harm to the environment."

OTHER COMMENTS

William Peck, supervisor of salvage, Admiralty, U.S. Navy, said the FOSC is unlikely to authorize a jettison unless laws or regulations provide specific provisions for doing so. Peck said he personally would not authorize a jettison, because "the law is so vague on this point and the potential liability so overwhelming."

Rear Admiral A.E. Henn, U.S. Coast Guard, concurred. "The chances of [the FOSC] making a decision to jettison are extremely slim, for all the reasons that we have stated so far."

Considerations in Making Time-Critical Decisions

Captain Kenneth J. Fullwood, manager of Maritime Relations, Environmental Affairs, Safety, and Nautical Services for Mobil Shipping and Transportation Co., outlined an accident scenario for two panels to discuss. An 80,000-deadweight-ton tanker was enroute from Mexico to the Delaware River with a cargo of 78,500 tons of crude oil. The captain has been plotting the track of a hurricane centered 430 miles away, just forward of the starboard beam and heading west at 12 knots. The storm was expected to cross the coast near Wilmington, North Carolina.

Due to a series of course changes and human errors, the ship ran aground off Cape Hatteras at 15 knots and came to rest with 80 percent of her length resting on soft sand. The bottom was not leaking and the power plant was not damaged, but the ship could not be backed off the shoal. It would take 16 hours to obtain a tug and barge from Norfolk to offload cargo, and 13 hours to get a light tug. The owner's technical experts and the classification society's naval architects determined that the vessel could not be refloated by internal cargo transfer. They also determined that discharging 2,000 tons of cargo would enable the vessel to back off the shoal, and furthermore, that unless the cargo was jettisoned within the next few hours, the ship would be driven further onto the shoal and break up as the winds and waves increased with the approaching storm.

Mr. Fullwood offered a decision tree diagram to assist in the decision-making process:

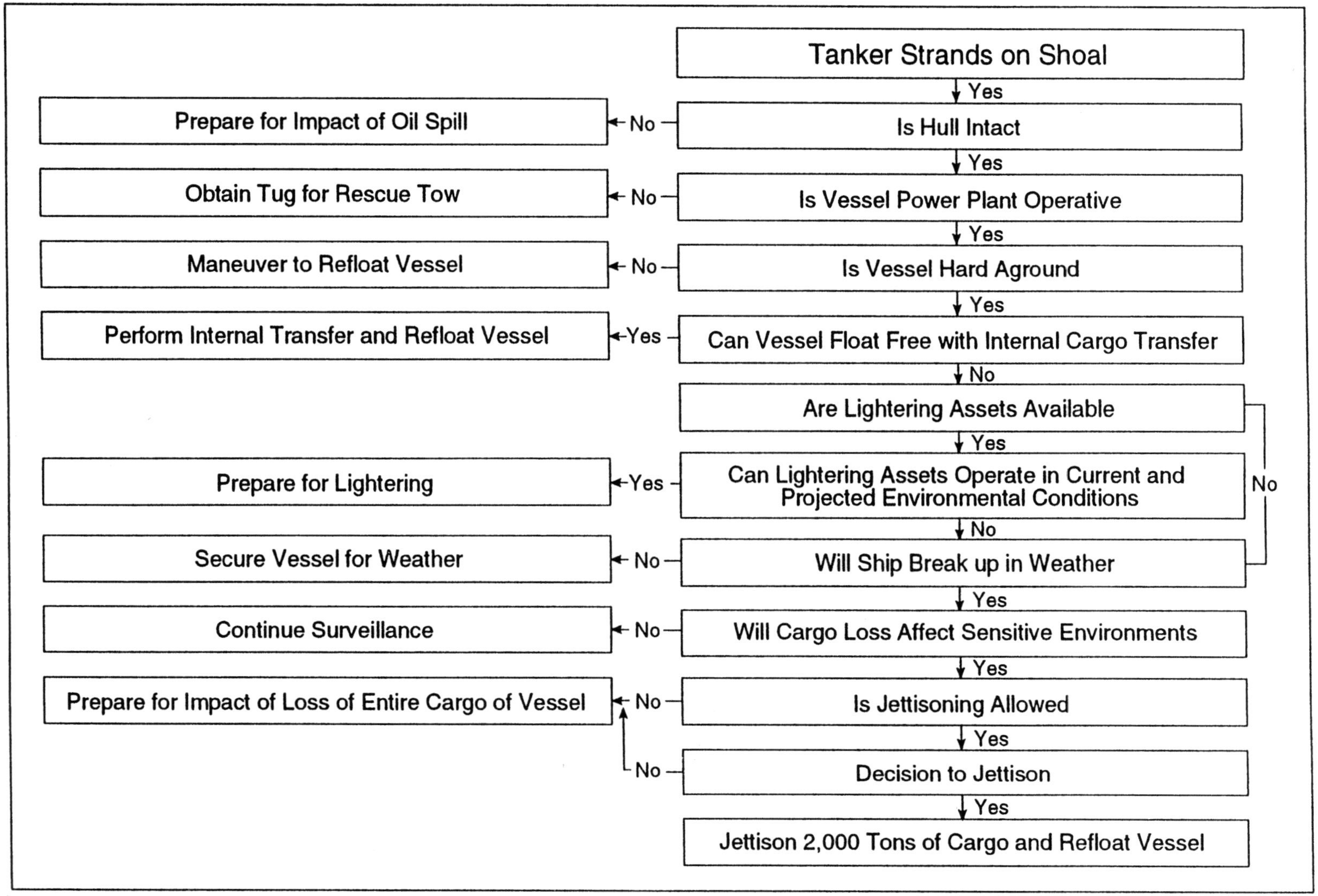

Decision Tree

Panel #1: Decision-Making Criteria

In the first panel, each participant explained his or her personal views regarding decisions whether or not to jettison oil.

Captain Donald S. Jensen, commanding officer of the National Strike Force Coordination Center based in Elizabeth City, North Carolina, described jettisoning as "a last resort to be employed only after all conventional alternatives have been carefully considered." If he were the FOSC in the scenario outlined, he would recommend a jettison of 2,000 tons of cargo to permit the vessel to refloat and get underway to escape the full impact of the hurricane. Jensen would recommend to his superior, the district commander, that the commandant authorize the master to jettison.

In making his decision, Jensen would consider the following criteria:

1. The time available for decision making.
2. Environmental forecasts, including tide and current patterns and projected weather.
3. Other response alternatives, such as lightering.
4. The trajectory of the 2,000-ton spill, based on a NOAA analysis.
5. Crude oil characteristics (the cargo is a light crude oil, much of which would evaporate quickly or be dispersed into the water column in the storm).
6. Shoreline impact and resources at risk, based on spill trajectory and vulnerability index maps.
7. Verification of vessel condition.
8. Verification of ship stability, in both damaged and intact condition, to confirm the amount of cargo to jettison.
9. Consultation of the area contingency plan.

Anne Rothe, the Alaska regional representative of the National Wildlife Federation, emphasized that the scenario reflects a failure of prevention and response mechanisms. She said the grounding occurred in "probably one of the most productive biological systems on the East Coast" and questioned why tankers were even allowed in that area. Furthermore, because Cape Hatteras is well known for its hazardous waters, she questioned why salvage and lightering equipment was not based nearby. More generally, Rothe expressed concern that any effort to facilitate jettisoning might discourage shipowners from securing lightering and salvage services. She also questioned the need to develop standard criteria for decision making, in that salvage situations vary so widely. She argued for retaining the current civil liability provisions in OPA 90, on the grounds they provide an effective prevention mechanism, but she added that responder immunity from criminal liability might be acceptable.

John Jay Driscoll, Jr., executive vice president of Smit Americas, said he doubted that 2,000 tons of jettisoned oil would be enough to free a vessel that grounded at 15 knots. He also observed that current liability risks almost prohibit jettisoning. Criteria used by Smit Americas salvage masters in assessing a casualty include:

- Circumstances of the stranding.
- Current situation of the vessel.
- External factors (e.g., tides, winds, current, traffic).

- Additional concerns regarding jettisoning, including trajectory of the spill and the character of the beach, water systems, and other impact areas.

The hurricane is the unpredictable factor in the scenario, thus meteorological information should be gathered quickly. In deciding whether to jettison, naval architects, the vessel's master, and the salvor should agree on whether the vessel is more likely to break up and spill the entire cargo than she is to survive, and whether the discharge of 2,000 tons of cargo would be enough to refloat the vessel. Smit Americas salvors probably would not jettison unless ordered to do so by the FOSC, Driscoll said, adding that the key issue then becomes who actually makes the final decision.

Roger Gale, manager of Shipping and Logistics for BP Oil Co., questioned whether sufficient tools and information are available in the scenario to make a jettison decision. Perhaps, for example, the ship's bottom is damaged more severely than it appears and would not hold up if the vessel were refloated. As it is not certain the ship would break up in the hurricane, which appears likely to veer north, Gale would not be prepared to jettison.

Jack Kalro, a naval architect with Diversified Technologies, outlined four conditions that should exist before jettison is recommended:

1. The vessel must be in imminent danger of breaking up. Given that the vessel in the scenario is on sandy bottom and grounded over 45 percent of her length, she appears to be fairly safe.
2. There is no other way of discharging cargo in the time available.
3. The jettison of a small amount of cargo would save the remaining cargo. In all likelihood, about 4,000 tons of oil would have to be discharged to save the 80,000-ton tanker.
4. All other possible solutions, however remote, must have been exhausted. More information would be needed before jettisoning in the scenario; a key factor is the liability involved.

Peter Bontadelli, administrator of California's Office of Oil Spill Prevention & Response and state on-scene coordinator for large spills, stressed that the common salvage practice of pumping water from flooded tanks raises the same issues as jettisoning. The biggest problem in a casualty situation is the availability of reliable information for decision making. Any vessel operating within U.S. waters must have a contingency plan that includes stability data and identifies nearby salvage capabilities and lightering vessels. Information also is needed concerning environmental risks to the area; unfortunately, most existing information is based on NOAA surveys that are at least 10 years old. A checklist should be available for use by on-scene coordinators in considering a possible jettison; a decision must be reached quickly, the needed information must be clearly available, and there must be a reasonable likelihood of refloating the vessel after the discharge. Bontadelli would not jettison under the scenario, because the likelihood of success is uncertain, and as a state official he has no clear-cut legal authority to do so. The NCP could be amended, with the consent of the Environmental Protection Agency and Coast Guard, to create the potential for such authorization, he said.

Barry Chambers of Clean America Inc. said all the decision-making criteria offered by participants were used in the *Argo Merchant* case, in which he was the salvage officer. According to Chambers, responders felt comfortable with the technical decision to jettison in that case, but the final decision was determined by social, economic, and political considerations. A pre-planned mechanism is needed for making jettison decisions.

Panel #2: A Regional Response Team's Decision-Making Exercise

The second panel discussed the information needed to make a decision about jettisoning and, as members of a mock regional response team, role-played a decision-making exercise for Captain Fullwood's scenario.

Mick Leitz, acting as salvage master, said he would not attempt to jettison without a tug attached, so the ship should be ballasted down and left to ride out the storm. Due to liability concerns, he would decline to accept any responsibility for the decision to jettison.

Captain Richard Fiske of the Navy said that based on technical input, jettison is an appropriate option to refloat the ship, legal concerns notwithstanding.

Philip Berns, representing the Justice Department, emphasized the need to consider possible environmental damage due to the discharge, adding that he would defer to technical experts.

Jerry Galt, providing scientific support, said the hurricane would cause a surge of two to six feet, resulting in a higher tide than predicted. Several hours would be needed to come up with a more precise figure. The high winds forecast would disperse the oil rapidly. The oil would reach shore probably within 12 hours and the sandy beach would be relatively easy to clean. In addition, the rain would help flush sensitive areas, and the reversal of the storm winds would help relax pressure from the oil in back marshes. If salvage experts were convinced the larger spill could be averted by the smaller one, then he would recommend a jettison.

Michael Ellis, representing the hull and machinery underwriters, said he would support a decision to jettison if the FOSC approved and if chances were good the discharge would permit the ship to be refloated.

Captain Fullwood, acting as the tanker owner's operating representative, said he would recommend a jettison, despite the risk of losing control over the refloated ship in strong winds. He emphasized that insurance would adequately cover the costs of pollution in the jettison case. Criminal prosecution was his main concern.

Warren Dean, as an attorney, said a foreign-flag operator without a U.S. office would be concerned principally with criminal liability for a spill. A U.S.-flag operator, on the other hand, would have unlimited liability under North Carolina law, even if a jettison were approved by the FOSC. The liability would be much greater for the full cargo than for the 2,000 tons. In any event, Dean said the decision maker should be indemnified against personal liability. He would recommend serious consideration of jettisoning, in spite of the ambiguities in state law and federal water pollution policy.

Nina Sankovitch, representing an environmental organization, emphasized the uncertainties in the scenario, including the condition of the hull, prospects for the vessel breaking up, and the path of the hurricane. Based on the salvage master's advice and the chance for averting any pollution, she advocated ballasting the ship and attempting to weather the storm. She questioned whether any legislation is needed to improve decision making.

Peter Bontadelli, representing the state, said all factors must be weighed, as they are before the use of oil dispersants. He gave odds of 3 or 4 to 1 that the damage could be reduced significantly by jettisoning. He would advise a jettison, despite the potential for liability under state law.

Mark Miller, of the National Response Corporation, said he could not respond to a spill immediately because the storm would place his crew and vessels in jeopardy.

Fred Burgess, representing the P&I Clubs[4], emphasized that any action taken should be directed by the Coast Guard. He would work with the vessel owner to assure positive public relations; response to many oil spills is driven by public opinion. Some mechanism is needed for public education, he said. Improvements in the decision-making framework could be done through regulation and the NCP, he suggested.

Jack Kalro, as the salvage engineer, would recommend a jettison.

Captain Don Jensen, acting as captain of the port and FOSC, noted that the decision would not be his alone, as the federal government advocates use of a Unified Command System (UCS) involving the FOSC, the state on-scene coordinator, and the vessel owner's designee.[5] He would be prepared to recommend to his superiors that a jettison be directed. However, Jensen doubted that an answer would be received in time to jettison under the scenario.

ANALYSIS OF KEY ISSUES

A number of themes and findings emerged from the symposium presentations, based on the committee's analysis.

Speakers generally agreed that jettisoning of oil can be a valuable salvage tool and should be considered as an option, to be undertaken only when failure to take such action probably would result in loss of the vessel and release of the entire cargo.[6] A deliberate discharge of a small volume of oil may be the only practical alternative in certain time-critical situations. Conventional alternatives such as lightering, or pulling grounded tankers off the shoals with tugs, may prove impossible due to the absence of appropriate assisting vessels.

Jettisoning has been rare in recent years. The speakers' varying interpretations of OPA 90 reflect the ambiguities in federal and state oil pollution laws and confusion within the maritime community concerning the legal effects of jettisoning. The Congress did not consider implications for salvage in developing OPA 90, and the resulting uncertainty over liability clearly is a factor in the reluctance to jettison.

It is critical to recognize that, under OPA 90, conditional immunity is not available to an individual who is grossly negligent or engages in willful misconduct. An intentional discharge would violate the FWPCA and most likely would be considered an act of "willful misconduct" by the courts, in effect eliminating the salvor's conditional liability and exposing the owner and operator to further liability. Furthermore, state laws may impose additional liabilities.

[4] A P&I Club is a group of shipowners who mutually agree to indemnify each other against amounts they are legally required to pay with respect to various types of liability incurred in the operation of their vessels, including liability for marine pollution, cargo damage, personal injury and death, and damage to piers, wharves, bridges, and other fixed structures.

[5] The UCS provides a framework for incident command decision making that ensures consultation and coordination among principal parties, in this instance the U.S. government acting through the federal on-scene coordinator; the state, acting through a predesignated representative, and the owner or other predesignated responsible individual. The UCS is a developing concept and structure.

[6] This thesis is supported by a 1993 resolution of the Maritime Law Association of the United States, which "supports the proposition that jettison of oil or hazardous substances should remain a viable option for ship masters and salvors if jettison may decrease the risk of loss of life or serious injuries or prevent discharge of greater amounts of oil or hazardous substances or more serious environmental consequences than the jettison itself."

The most direct means of increasing liability protection for salvors and other responders may be to amend the NCP to clarify the procedure for arriving at a decision to jettison and to place the responsibility solely on the FOSC. Such an approach would obviate the need to persuade the Congress to amend OPA 90, or to await a judicial interpretation following an incident of jettisoning. This change would not solve the problem fully, however, because OPA 90 expressly does not preempt state law, and the salvor may be exposed to additional liability directly or indirectly under general maritime law or various state laws.[7] In any case, clarification of oil pollution laws undoubtedly will require further judicial or regulatory interpretations and, at least in some instances, further legislation.

Participants differed as to whether a jettison is an appropriate response to Captain Fullwood's scenario. This disagreement demonstrates the difficulty and subjective nature of such decisions and suggests a need for standard, objective decision-making criteria. Such criteria could help expedite a process that involves multiple decision makers and special interests. The following criteria were suggested as fundamental conditions that must exist before any oil is jettisoned:

- Time pressures demand immediate action.
- Deliberate discharge of the proposed amount of oil is likely to save the ship and the remaining cargo.
- All other salvage options, such as internal cargo transfer and lightering, have been exhausted or considered and rejected.
- Failure to jettison is likely to lead to loss of the ship and release of the remaining cargo. The principal issue is whether the ship will break up in bad weather, so information is needed concerning tides, currents, and approaching storms.
- The condition of the stranded vessel – her hull and her intact or damage stability – is adequate so that the ship could be refloated following the jettison, and the remaining cargo saved.
- All necessary preparations have been made, including the marshaling of tugs, if available, to refloat the ship quickly after the discharge.
- The FOSC is monitoring the situation continuously to ensure that jettisoning remains the only viable option.
- Preparations are underway to clean up the discharged oil. Information is needed concerning spill trajectory, characteristics of the oil, physical environmental conditions, containment and recovery measures, geology of the impact zones, toxicological sensitivity of vulnerable species, and ecological characteristics of vulnerable areas.

These conditions aside, the committee recognizes that there may be instances when the lives of the crew and passengers (if any) may be endangered by the breaking up or sinking of a vessel. In such a case, if loss of the vessel and the consequent danger of loss of life could be avoided by the discharge of all or part of the cargo, the mariner concerned about survival would order the discharge, no matter where it was

[7] States have no jurisdiction (for this purpose) beyond the three-mile territorial sea. States do assess penalties and attach liability, however, to those responsible for spills that originate outside the territorial sea and drift into it.

made, even though there might not be time to request and obtain permission from authorities.

Other comments at the symposium indicated that the time-critical information needed when considering a jettison may be provided by computer modeling programs. Current models are capable of estimating currents and the distribution of jettisoned oil over a 24-to-36-hour period. It seems clear that these models can play a useful role in clarifying alternative courses of action in the decision-making process – provided the input data are sound and calculations can be made available to responders in a timely manner.

Finally, two general factors that may impede sound salvage practices were mentioned. Several speakers indicated that, even if jettisoning appears to be the correct technical decision, the FOSC in the decision-making exercise only recommends this action to superiors – first the district commander and, ultimately, the commandant. This places the issue in the political arena, as occurred in the *Argo Merchant* case. Under these circumstances, and without specific criteria on which to base a decision to jettison, public environmental concerns effectively may block action.

The other factor is the uncertain legality of discharges that may occur during the normal course of salvage. A number of tools traditionally employed by salvors could be deemed a form of jettisoning, as they may result in a discernible discharge of oil. Examples include pumping out a flooded engine room, pressing down of dirty ballast tanks, expelling water from a flooded cargo or fuel tank, using compressed air to press out damaged tanks, displacing oily water with buoyant material, and operating many on-water skimmers (which, in separating oil and water, may discharge small quantities of oil). Such actions result in minimal pollution and likely would be part of an approved plan of action; yet, regardless of their benefit, these incidental discharges may violate the FWPCA.

CONCLUSIONS AND RECOMMENDATIONS OF THE COMMITTEE ON MARINE SALVAGE ISSUES

Jettisoning of petroleum cargo can be a valuable salvage tool and should be considered as an option, to be undertaken only when failure to take such action might, and probably would result in loss of the stranded vessel and release of the entire cargo. However, a number of unresolved issues tend to inhibit the reasoned use of jettison.

In the committee's judgment, these issues should be resolved before a marine casualty occurs where a deliberate discharge of oil may be warranted. Otherwise, in the absence of legal certainty, a salvor may reject the jettison option arbitrarily, even when it may be the only means available to avoid a catastrophic spill.

The committee concludes that some unresolved issues regarding intentional discharges could be clarified expeditiously by amending the NCP, and that so doing might avert serious liability problems without requiring any changes in OPA 90. OPA 90 provides conditional immunity to persons acting in the course of rendering care, assistance, or advice that is consistent with the NCP. The committee concludes that, at present, it is unclear whether a jettison may be considered consistent with the NCP. The committee therefore recommends:

The NCP should be amended to accomplish the following objectives: To give the FOSC explicit authority, in consultation with the appropriate state authority, to approve the jettison of a situation-specific amount of oil under certain limited circumstances, to save a vessel and those on board, as well as

her remaining cargo; and to provide procedures whereby such action may be authorized and undertaken.

Such an amendment would resolve much of the uncertainty as to salvor liability and immunity by making the act of jettisoning an authorized and viable option in response to the threat of a catastrophic oil spill. This change also would limit owner and operator liability for the salvor's actions. Such an amendment probably would not protect a salvor against liability under state laws. However, most state statutes require either consistency with the NCP or at least the absence of conflict. Therefore, if in addition to consulting with the states as required by federal law, the FOSC obtains concurrence from the state incident commander consistent with the NCP, there is a possibility that immunity also could be obtained under state laws.

A related issue concerns certain common salvage practices that also could be considered forms of jettisoning, in that some oil may be discharged. These actions include pumping out a flooded engine room, pressing down of dirty ballast tanks, expelling water from a flooded cargo or fuel tank, using compressed air to press out damaged tanks, displacing oily water with buoyant material, and operating on-water skimmers. The committee concludes that a salvor should be afforded protection to use these tools under certain limited conditions. The committee therefore recommends:

The NCP should be amended to give the FOSC explicit authority, in consultation with the appropriate state authority, to approve certain common salvage actions that may result in incidental discharges of small quantities of oil. Such actions include pumping out a flooded engine room, pressing down of dirty ballast tanks, expelling water from a flooded cargo or fuel tank, using compressed air to press out damaged tanks, displacing oily water with buoyant material, and operating on-water skimmers. The FOSC authority could be contained in approval of the daily work plan, which, if carried out under the UCS, also could be approved by the state.

The committee further concludes that the present lack of official, objective criteria for reaching a technical decision to jettison oil may undermine decision making during salvage situations. Such decision making needs to be logical, timely, reliable, and defensible, and the FOSC needs to be diligent in analyzing the relevant issues. The committee therefore recommends:

The Coast Guard should develop a checklist containing specific conditions that must be met as prerequisites for a decision to jettison oil. The FOSC should follow the checklist in authorizing such action under the NCP. Responder conformance with the checklist and with an FOSC decision authorizing the jettisoning of cargo should ensure full protection against liability for a salvor who jettisons oil.

Such a checklist might include the following criteria:

- Time pressures demand immediate action.
- Discharge of the proposed amount of oil is likely to save the ship and the remaining cargo.
- All other salvage options, such as internal cargo transfer and lightering, have been exhausted.
- Failure to jettison is likely to lead to loss of the ship and the remaining cargo.
- The condition of the stranded vessel is adequate so that the ship probably can be refloated and the remaining cargo saved.

- All necessary preparations have been made, including the marshaling of tugs, if available and needed, to refloat the ship quickly.
- The FOSC is monitoring the situation continuously to ensure that jettisoning remains the only viable option.
- Preparations are underway to clean up the discharged oil.

An amendment to the NCP establishing the process, standards, and criteria for authorizing a jettison or similar discharge would be consistent with the President's authority to direct removal actions as provided under existing law. Specifying conditions when jettisoning may be carried out would indicate clear "direction" from the President and would advance the congressional intent to facilitate prompt and effective response.

Explicit authorization for the act of jettisoning also could enable the salvor to avoid criminal or civil penalties that otherwise might be imposed for an unauthorized discharge. Moreover, under certain circumstances, the salvor would be immune from liability for removal costs or damages resulting from the jettison, because these actions would be both consistent with the NCP and undertaken at the direction of the FOSC acting for the President. Utilization of the UCS and the receipt of state concurrence in a decision to jettison also could provide protection in some states.

Many issues would remain unresolved, however. In particular, it is not clear whether the states could impose their own criminal or civil penalties on a salvor who jettisons into state waters, and, if not, whether the states could impose liability on the responsible party for damages resulting from the salvor's act. Utilization of the UCS, and modifications to state contingency plans in line with the committee's recommended changes to the NCP, may help resolve some of these issues.

Other questions concern whether a responsible party may seek general contribution or indemnification for such acts, or whether certain general maritime law claims and defenses exist; and how the 1851 Limitation of Liability Act applies in light of OPA 90. Implementation of the committee's recommendations to clarify both the authorization and the criteria for jettisoning should go a long way toward resolving these issues, if and when the occasion arises.

APPENDIX C

RESPONDENTS TO QUESTIONNAIRE

Alfonso M. Sotres and Associates, Ltd.
American Hull Insurance Syndicate
American President Lines
ARCO Marine, Inc.
Atlantic Container Lines
Clean America Environmental Services
Clean Sound Cooperative
Cook Inlet Regional Citizens Advisory Committee
Crowley Marine Services, Inc.
Diversified Technologies
Foss Maritime Co.
Haight, Gardner, Poor, & Havens
International Chamber of Shipping
MacKinnon Searle Consortium, Ltd.
Marine Pollution Control
Maritrans
Marpol, Incorporated
McCallister Towing of Wilmington, North Carolina
Milwee Associates, Inc.
Mobil Oil
MSD Morgan City
Marine Safety Office, Alameda
Marine Safety Office, Boston
Marine Safety Office, Cleveland
Marine Safety Office, Corpus Christi
Marine Safety Office, Galveston
Marine Safety Office, Hampton Roads
Marine Safety Office, Honolulu
Marine Safety Office, Houston
Marine Safety Office, Jacksonville
Marine Safety Office, Juneau
Marine Safety Office, Port Arthur
Marine Safety Office, Portland
Marine Safety Office, Puget Sound
Marine Safety Office, San Diego
Marine Safety Office, San Francisco
Marine Safety Office, Savannah
National Strike Force Coordination Center
Pen Attransco Corp
Sea Brex Marine
SKAARUP Oil Corporation
Smit Americas
State of California Department of Fish and Game

Stickney, Dufour & Associates, Inc.
The Glosten Associates
The Salvage Association, New Orleans
The Salvage Association, New York
The Salvage Association, San Francisco
Titan Maritime Industries, Inc.
Turecamo of Savannah, Inc.
W.H. Padies Associates, Inc.
Wijsmuller Salvage USA, Inc.

APPENDIX D

INVITED PARTICIPANTS AT COMMITTEE MEETINGS

COMMITTEE ON MARINE SALVAGE ISSUES
Washington, D.C.
August 11-12, 1992

Joseph Cox, American Institute of Merchant Shipping
Laurie Crick, Dyer, Ellis, Joseph & Mills
Michael Donohoe, Marine Environmental Protection Division, U.S. Coast Guard
William Duncan, Exxon Shipping Co.
Richard P. Fiske, Ocean Engineering/Salvage and Diving, Naval Sea Systems Command, U.S. Navy
Richard Fredericks, Marine Spill Response Corporation
Jack McGrath, Marine Spill Response Corporation
Paul B. Mentz, Office of Technology Assessment, U.S. Maritime Administration
Willard F. Searle, Jr., NAE, MacKinnon Searle Consortium, Ltd.
Don Walsh, Marine Board and International Maritime, Inc.
Kerry Walsh, Fred Devine Diving and Salvage, Inc.

SYMPOSIUM ON THE PURPOSEFUL JETTISON OF CARGO
Washington, D.C.
February 23, 1993

Thomas Allegretti, American Waterways Operators
Tim M. Beaver, Global Diving and Salvage
Alan Becker, PCCI
Philip Berns, Department of Justice
Jim Bladh, U.S. Navy Naval Sea Systems Command Office of Salvage and Diving
Alex Blanton, Dyer, Ellis, Joseph & Mills
Liz Sigel Bouchard, Vessel Operations and Maritime Initiatives, Transportation Institute
Carolann Bowen, Office of Coastal Protection, Florida Department of Natural Resources
Alan Breed, Marine Spill Response Corporation
Richard H. Brown, Kirlin, Campbell, Meadows & Keating
Fred Burgess, Le Boeuf, Lamb, Leiby, and Macrae
Robert T. Bush, Universe Tankships, Inc.
Edward G. Cawthon, Turecamo of Savannah, Inc.
Barry E. Chambers, Clean America, Inc.
William Chubb, U.S. Coast Guard
Kevin Cohen, Dyer, Ellis, Joseph & Mills
Mark Cohen, Royston, Rayzor, Vickery & Williams, L.L.P.

Patricia A. Collins, U.S. Coast Guard National Pollution Funds Center
Sean T. Connaughton, Haight, Gardner, Poor & Havens
Philip Cooney, American Petroleum Institute
John D. Costello, Marine Spill Response Corporation
Joseph J. Cox, American Institute of Merchant Shipping
Laurie Crick, Dyer, Ellis, Joseph & Mills
Lee R. Crockett, Committee on Merchant Marine and Fisheries Staff, U.S. House of Representatives
Thomas F. Daly, McCarter & English, Esquires
Andrew W. D'Angelo, MacKinnon Searle Consortium Ltd.
Warren Dean, Dyer, Ellis, Joseph & Mills
Graham H. Deere, The Salvage Association
Rita J. Diehl, Council on Ocean Law
Michael Donohoe, U.S. Coast Guard Environmental Protection Division
J. Stephen Dorrier, Marine Spill Response Corporation
John J. Driscoll, Smit Americas, Inc.
Ron Duckhorn, Crowley Marine Services
William Eichbaum, Marine Board and World Wildlife Fund
Michael Ellis, The Salvage Association
Rainer Engelhardt, Marine Spill Response Corporation
Glenn Epler, U.S. Coast Guard
Gary Faber, Crowley Marine Services
Richard B. Fairbanks, Titan Maritime Industries, Inc.
Bruce Fernie, Fleet Management, Texaco Marine Services, Inc.
Richard P. Fiske, U.S. Navy Naval Sea Systems Command Office of Salvage and Diving
Richard Fredricks, Marine Spill Response Corporation
Roger Gale, BP North America
Jerry Galt, National Oceanic and Atmospheric Administration
William O. Gray, Skaarup Oil Corporation
David Hall
Paul Hankins, U.S. Navy Naval Sea Systems Command Office of Salvage and Diving
Robert E. Hartzel, Ellsworth Salvage, Inc.
John R. Henley, MacKinnon Searle Consortium Ltd.
A. E. Henn, U.S. Coast Guard
David V. Hutchinson, U.S. Department of Justice
Donald S. Jensen, U.S. Coast Guard National Strike Force Coordination Center
Marcus J. Johnson, Consultant
Jack Kalro, Diversified Technologies
Eugene M. Kelly, Amoco Transport Company
Christopher Kende, Holtzmann, Wise & Shepard
David M. Kennedy, National Oceanic and Atmospheric Administration, Hazardous Materials Response and Assessment Division
Richard Knee, U.S. Coast Guard
John Koster, Healy and Baillie
Miles Kulukundis, INTERTANKO, United Kingdom P & I Club, Salvage Working Group
J. T. Leigh, Marine Spill Response Corporation
Marilyn Leland, RCAC
Malcolm MacKinnon, MacKinnon Searle Consortium Ltd.
Charles S. Maclin, Gallagher Marine Systems, Inc.
Gerald A. Malia, Kirlin, Campbell, Meadows & Keating
Henry Marcus, Massachusetts Institute of Technology
Bruce B. McCloskey, MacKinnon Searle Consortium Ltd.

James McDonald, Marine Spill Response Corporation
Jack McGrath, Marine Spill Response Corporation
William Merlin, Marine Spill Response Corporation
Trygve A. Meyer, INTERTANKO
Mark Miller, National Response Corporation
W. I. Milwee, Milwee Associates, Inc.
Jim Morgan, Arco Marine
Robert H. Nicholas, Jr., Exxon Company, U.S.A.
Barry Ray Ogilby, Marine Spill Response Corporation
Skip Onstad, Marine Spill Response Corporation
Hans Kristian Øvstaas, The Bergesen DY Group
Paul Preus, MacKinnon Searle Consortium Ltd.
Albert A. Propsma, Smit Americas, Inc.
V. Lee Okarma Rees, Graham & Dunn
Klaas J. Reinigert, International Salvage Union and SMIT TAK B.V.
Kent Roberts, Schwabe, Williamson and Wyatt
Richard Roth, Crowley Marine Services
Anne Rothe, National Wildlife Federation
John R. Sambrook, The Salvage Association
Andrew Santos, Cook Inlet RCAC
W. F. Searle, MacKinnon Searle Consortium Ltd. (retired)
David J. Sharpe, National Law Center, George Washington University
Sidney H. Shaw, MacKinnon Searle Consortium Ltd.
Mike Sowby, Office of Oil Spill Prevention and Response, California Department of Fish and Game
Malcolm L. Spaulding, University of Rhode Island Department of Ocean Engineering
Rosemary Stein, Exxon Shipping Co.
James Stillwaggon, MacKinnon Searle Consortium Ltd.
John L. Sullivan, Smit Americas, Inc.
Robert L. Sullivan, American Petroleum Institute
R. K. Thurman, MacKinnon Searle Consortium Ltd.
Gene Toffoli, Office of Oil Spill Prevention and Response, California Department of Fish and Game
Andreas K. L. Ugland, INTERTANKO
David Usher, Spill Control Association of America
George L. Waddell, Hancock, Rothert & Bunshoft
John Waldron, Marine Spill Response Corporation
Sidney A. Wallace, Dyer, Ellis, Joseph & Mills
Ritner E. Walling, Ellsworth Salvage, Inc.
William Webster, University of California at Berkeley and Commission on Engineering and Technical Systems
Russell Weil, Kirlin, Campbell, Meadows and Keating
Edson Whitaker, MacKinnon Searle Consortium Ltd.
David L. Wood, Health, Safety and Environmental Protection Department, Marine Transport Lines, Inc.
Frank Wood, U.S. Coast Guard Office of Marine Safety, Security and Environmental Protection
John M. Woods, Thacher, Profitt and Wood

REGIONAL ASSESSMENT
OF MARINE SALVAGE CAPABILITIES
Seattle, Washington
June 3-4, 1993

J. Tom Bringloe, Glosten Associates
Bill Carey, U.S. Coast Guard Marine Safety Office, Puget Sound
Ron M. Duckhorn, Crowley Marine Services, Inc.
Jim Dunlap
John Dwyer, U.S. Coast Guard Marine Safety Office, Puget Sound
Gary C. Faber, Crowley Marine Services, Inc.
Cdr. Hall, USCG Dist.
Nick Handy, Washington State Office of Marine Safety
Harlan Henderson, U.S. Coast Guard
John Keenan, Sea-Land Services, Inc.
Robert Levine, Prince William Sound Towing Study
Ron Lokites, U.S. Coast Guard Marine Safety Office, Puget Sound
William I. Milwee, Milwee Associates, Inc.
James M. Morgan, Arco Marine, Inc.
Roger Mowery, U.S. Coast Guard Marine Safety Office, Puget Sound
Don Patterson, Columbia Helicopters, Inc.
Paul Poliak, Maritime Law Association Subcommittee on Salvage
Denny Quirk, Clean Sound Cooperative, Inc.
Richard Roth, Crowley Marine Services, Inc.
Steve T. Scalzo, Foss Maritime Company
Paul Slyman, Oregon Department of Environmental Quality
George L. Stiehl, West Coast Shipping Company
Ben Strickland, Crowley Marine Services, Inc.
Kerry Walsh, Fred Devine Diving and Salvage

REGIONAL ASSESSMENT
OF MARINE SALVAGE CAPABILITIES
ON THE EAST COAST OF THE UNITED STATES
New York, New York
August 3, 1993

Gerald W. Abrams, U.S. Coast Guard Marine Safety Office, Boston
Bruce Banks, Jamestown Marine Services
Peter Becker, Aetna
George C. Blake, Maritime Overseas Corporation
Michael Butler, New York City Fire Department Marine Division
Graham Deere, The Salvage Association
J. Kenneth Edgar, Diversified Technologies
Andre Galerne, IUC
Jack Geck, California Office of Oil Spill Prevention and Response
Neil Hansen, Jamestown Marine Services
John Johnson, Bouchard Transportation
Mark H. Johnson, U.S. Coast Guard Marine Safety Office
Malcolm MacKinnon, MacKinnon Searle Consortium Ltd.
Davella May, Port Authority of New York and New Jersey

John Ringelberg, Jamestown Marine Services
Rodney Sambrook, The Salvage Association
James T. Shirley, Haight, Gardner, Poor and Havens
Donna Siracusa, U.S. Coast Guard
James P. Sweeney, Morania Oil Tanker Corporation
Ritner Walling, Ellsworth Salvage
George Wittich, Weeks Marine

REGIONAL ASSESSMENT OF MARINE SALVAGE CAPABILITIES ON THE GULF COAST OF THE UNITED STATES
New Orleans, Louisiana
August 9, 1993

Joe Ahern, Shell
James C. Arnold, Aramco Services Company
Cappy Bisso, Bisso Marine
Captain Calhoun, Marine Safety Detachment, Eighth District, U.S. Coast Guard
Stephen Dick, Tidewater Marine
John Driscoll, Smit Americas, Inc.
Richard B. Fairbanks, Titan Maritime Industries, Inc.
Jack Geck, California Office of Oil Spill Prevention and Response
Doug Lentseh, U.S. Coast Guard, New Orleans
Paul Manzi, Sabine Transportation Company
Cary McNamara, U.S. Coast Guard Marine Safety Office, New Orleans
Jim Obernesser, U.S. Coast Guard Captain of the Port, New Orleans, Port Operations
Steve Olson, U.S. Coast Guard Marine Safety Office, New Orleans
Ted Thompson, U.S. Coast Guard Captain of the Port, New Orleans
Roger White, Edison Chouest Offshore
Les Williams, Williams Fire and Hazard Control
Randi Williams, Williams Fire and Hazard Control

APPENDIX E

INSTANCES OF SALVAGE IN U.S. WATERS REPORTED TO THE COMMITTEE[1]

Vessel Name	Vessel Type	Situation	Response	Critical Factors	Outcome
	Hess Oil storage tank	Ruptured; 185,000-barrel crude oil spill	Fast	Fast response	90% of oil recovered; small amount of oil got into the Kill van Kull
		9,500,000-gallon waste oil spill	More than 200 men working each day with various equipment for 5 weeks		
	Russian trawler	Fire	USCG directed response by Crowley and Boots and Coots under Navy contract		Vessel returned to service
	Drilling rig	Adrift in heavy weather	Tow	Weather; communication; crew fatigue; equipment failure; lack of specific equipment	Rig sank under tow
	Barge/tug	Barge parted tow-line; could not be boarded by crew of tug		Weather; communication; fatigue of crews; equipment failure; lack of specific kind of equipment	
	Deep notch barge	Collision	Secure damaged fuel piping; make fuel system operable to aftermost pumping engine; relocated cargo hoses to enable overhead pumping; lightered approx. 500,000 gallons of gasoline		
	Barge	Grounded	Offloaded cargo from undamaged compartments; barge refloated after being pulled from rocks by 2 tugs; ballasted to stop further leaking	Gasoline leaking from all damaged tanks; gale warning	
A. REGINA/ 3EKM	M/V passenger ferry	Aground	LOF SMIT/SASI-liaison		Vessel lost, given up
ABIQUA/ KTZO	Bulk motor		Partial offload; tow condition		
ADMIRALTY BAY	M/V		Dry salvage		
AEGEAN SUN	M/T		Dry salvage		
AFRICAN PIONEER	M/V general cargo	Collision	Flooded engine room; towed to Galveston, TX; engine room pumped dry; towed to Brownsville, TX; broken up; SASI/SUSTA		Broken up

[1] The committee was unable to verify the facts of the following incidents; the information is reported exactly as received.

Vessel Name	Vessel Type	Situation	Response	Critical Factors	Outcome
AIKATERINA	Tankship	Fire	Fire extinguished by USCG and Boots and Coots personnel after burning for several days	Dealing with offshore vessel fires	Vessel not lost at sea
AKRON/ D5VF	M/V bulk cargo	Aground	SASI refloated; reload		
ALEXANDERS ABILITY		Aground; engine room flooded			
ALL ALASKAN	Fish processor	Grounded	Oil removed by USCG; fish processing machinery salvaged by owner; unsuitable conditions for salvage	Broached and broke while tow line being cast	No assets
ALVENUS	MV tanker	Aground	Lightering operation offshore; antipollution operation a.o. Galveston Beach		Major oil spill
AMOCO CADIZ	VLCC	Loss of steering; grounding	Salvage tug on scene within 3 hours; unsuccessfully attempted to take AMOCO CADIZ under tow	Main engine functioning but not used during attempts to establish a tow	Major pollution occurred; ship total loss
AOYAGI MARU	F/V	Grounded	USCG burned remaining fuel using explosives	Owners looked at limits of liability vs. salvage value	No salvage conducted
APEX 3405	Tank barge	Collision; sank	Raised sunken barge; vessel removed from channel	Channel closure restrictions; no need to maintain intact cargo envelope; stability calculations not necessary	Discharged over 750,000 gallons of catalytic feedstock oil
AQUILA AZTECA/ XCGU	M/Tank	Grounded	Partly offloaded; refloated; SMIT/SASI		
ARCO ANCHORAGE	T/V	Grounded; cargo tanks punctured	USCG and commercial submersible pumps used to transfer crude oil to lightering barge; divers put in temporary patch; vessel sailed to Cherry Point, WA; offloaded cargo		Went to shipyard in Portland, OR
ARGENTINA	Ocean-going ferry	Grounded	National Strike Force personnel pumped fuel oil from vessel; vessel was cut down to the waterline and superstructure taken away	Remote location; environmental sensitivity of the area	
ARGO IDUNA		Explosion in engine room	Lightering operation		
ARGO MERCHANT	Handy size tanker	Grounding	MPS ship CURB on scene; lightering available and USCG ADAPTS pumps on board, but lack of steam generator prevented heating oil for offloading when weather permitted	Winter; MPS ship CURB in Key West, took 5 days to arrive on scene; lack of steam generator prevented heating oil for offloading; divided authority on scene (USCG, State of MA, EPA)	Weather worsened; ship broke up; ship and cargo total loss
ARROW	Coastal tanker	Grounded, then sank in 60 feet of water	Cargo offloaded using first-time application of hot-tap technique; local heating at off loading pumps was necessary; heating process very inefficient	Winter	No pollution occurred
ATLANTIA			Pre-inspection; installation		
ATT STALL	M/T		Wet salvage		
B 155	Barge	See OCEAN 255			
BANGLER BANNI	M/V		Dry salvage		

Vessel Name	Vessel Type	Situation	Response	Critical Factors	Outcome
BARGE 160-4; AVENGER	1,600-DWT tank barge; tug	Grounded	Salvage team; fly in equipment	Barge empty; barge sustained only moderate bottom damage; tug sustained extensive bottom and side shell damage	Fuel removed from tug; both refloated by use of ground tackle
BARGE 500-4	12,000-DWT rail car and trailer barge	Grounded	Damage control efforts to limit flooding by pressurizing selected tanks; use of properly equipped salvage vessel (M/V ARCTIC SALVOR) tugs	Full load of cargo; extensive damage to starboard-side shell and bottom-plating; use of properly equipped salvage vessel and tugs with both power and sea-keeping ability to work in prevailing weather conditions	Successfully refloated without loss of cargo
BARGE BEAN 12			Wet salvage		
BARGE KENAI	9,750-dwt deck barge	Grounded	Salvage team on-site within 6 hours; utilized salvage vessel ARCTIC SALVOR	Extensive bottom damage and subsequent flooding; prompt arrival of salvage personnel; availability of specialized equipment capable of being flown to the site; delays were caused by high wind and adverse sea conditions	Successfully refloated without loss of cargo
BEAN 12	Anchor handling vessel	Sunk	Removal with Derrick barge		
BETTY L	Deck barge	Aground; USCG removed 120,000 gallons of diesel fuel	Salvage chief removed barge successfully		
BLUE MAGPIE	Cargo vessel	Grounded on breakwater		Timely availability of salvage assets	Broke up before salvage assistance could reach scene
BRAZILIAN FRIENDSHIP		Grounded; broken rudder	Refloated with local salvor and tugs; used tugs to steer vessel down river to awaiting foreign tug (BUGSIER) towed to Germany	Fully loaded iron ore	
BT 155	Barge	See OCEAN 255			
BT NAUTILUS	Barge	Grounded in Kill van Kull		Partial closure of waterway for several days	
BURMAH AGATE / ELZS	M/Tank	Collision; fire			Vessel burnt out; 19,000 tons crude saved
BYRON 16		Aground	Cargo lightered with severe loss; damaged in refloating	Poorly chosen salvage equipment	
C/V STAR LEKINGER		Grounded on aft third	International salvage firm hired by responsible party at the direction of COTP; acquired 3 high HP tugs; had vessel off-ground within 36 hours	Greater than normal east-west current continued to set the vessel into shallow water; fear of possibility of double bottom fuel tanks rupturing and releasing bunkers into the sea; miles of pristine white beaches in direct path of potential fuel spill; no tugs of sufficient horsepower locally available	No pollution resulted; no damage other than scraped paint and slight indentation along starboard quart

Vessel Name	Vessel Type	Situation	Response	Critical Factors	Outcome
CAMARGUE / FNUB	M/T	Collision in ballast tank	Vessel inerted; damage covered; assisted into Galveston SASI		
CARIBE	Barge	Aground			
CC-11	Barge	Stranded and sank		Owner failed to employ competent salvors	Vessel broke up and became wreck removal project
CEPHEUS	T/V	Grounded	Refloated itself; headed for port with damage to double bottom and cargo tank; steam trace rigged to remove heavy fuel from double bottoms; offloaded cargo		Sailed to Kachemak Bay for survey and repairs
CHARA-LAMBOS	Large tanker	Large section of side plate fell off while being towed to scrap	Flooded opposite side tanks sufficiently to bring upright	Time critical to even up and bring open section of hull clear of water to prevent extension of opening	Continued in tow
CHEM 2	Tankbarge	Stranded on sandbar	Equipment and trained ANH personnel brought in; dredged channel to barge; lightered cargo; winched tankbarge and lightering barge out	Silting; rapid river stage fall; poor access; downwind community; potential loss of cargo cooling	
CHEVRON GUAM	Small POL tanker	Grounding in typhoon	Search and survey in deep water where tanker was left by forces of nature	Full load of avgas and jet fuel	Ship was total loss; no pollution due to volatile nature of cargo and scrubbing forces of nature
CHEVRON HORIZON	Crude oil tanker	Grounded and holed; hit in stern by gasoline tow	Tanker refloated with tugs; ballasted down creating water bottoms; temporary patch by divers; cargo offloaded		Sent to drydock
CHINA PRIDE	Ore carrier	Stranded on sandbar	Lightered and pushed off with tugs	Strong current; dropping river stage; silting	
CIUDAD DE BRAQUE			Wet salvage		
CIUDAD DE IBAGUE	M/V	Holes in watertight bulkhead	Hull plate removed		
CIUDAD DE PASTO		Hull fracture; taking water in machinery space and three holds	Determine location of vessels and equipment; hire tug and supply vessel; establish towage connection	Bad weather; time critical situation because sinking of the vessel was threatened; crew had abandoned vessel; only vessel's emergency generator was available to supply power	
COLUMBIA NEW YORK	Barge	Grounded	Refloated with tug assist	Time critical as bad weather could have broken her on rocks	
CRANE RECOVERY			Wet salvage		
CYRIL	M/V general cargo	Aground	Offload; refloat		
DAE RIM	Fishing vessel	Drifted ashore after unsuccessful rescue tow		Failure caused by Russian tug letting go her tow	
DAY ISLAND	Fishing vessel	Swamped over underwater rock	Refloated		Docked
DELTA AMERICA/ KFFJ	M/V general cargo	Aground	LOF Moran-Crowley; SMIT/SASI		

Vessel Name	Vessel Type	Situation	Response	Critical Factors	Outcome
DEVALI 1/ D51W	M/Tank	Aground during hurricane	Inspection and contract LOF; SMIT	Loaded crude	
DIALA	MV tanker	Aground	Lightering operation "over the top"		
DIALA/ DGBS	M/Tank	Aground	Partial discharge; refloated; SMIT/OCEAN - salvage	Cargo crude	
DIAZ ORDAZ	M/F		Wet salvage		
DIMI		Aground			
DOCELOTUS/ PPVD	M/V general cargo	Aground	Refloated after deballasting		
DREDGE BARGE	Barge		Survey		
DUVAL II	Barge	Collision and subsequent sinking in Houston Ship Channel	Pushed to side of channel; heavy lift barge from New Orleans salvaged vessel	Unavailability of heavy lift barge in area; h-l barge delayed by fog	Successfully salved; two way traffic not restored for about two weeks
EAGEAN SUN	MV tanker	Struck submerged object and grounded	Inerted all cargo tanks and refloated vessel		
EASTERN STAR	Log carrier	Broke mooring	Controlled by local harbor tug	Capable vessel willing to help	Continued voyage
EL PASO PAUL KEYSER	LNG tanker	Grounding	Vessel dispatched for lightering; special offload equipment mobilized immediately by owner;	Double bottom prevented penetration of inner hull; ship provided with and had trained in damage control and casualty procedures; owners on scene within 30 hours to liaise with local authorities and government officials and prepared to make tough decisions including destruction of ship; salvage engineers and naval architects on scene, backed by support in CONUS	No pollution and no litigation occurred; ship repaired in drydock and back in service
ELDIAM/V		Survey			
EMS ORE		Grounded; broken rudder	Refloated with local salvor and tugs; used tugs to steer vessel down river to awaiting foreign tug (BUGSIER) towed to Germany	Fully loaded iron ore	
ENTERPRISE	MV bulkcarrier	Aground	Lightering operation		
ENTERPRISE /3FOF	M/V bulk	Aground	Partial offload; refloat SASI; delivered to Beaumont, TX; SASI		
ERKOWIT	Breakbulk	Collision with DORTMUND	Entry into inner harbor denied; ship was sinking; grounded by salvage master on local public beach; hazardous deck cargo floated off	Deck cargo of highly toxic agricultural chemicals; tug master doubled as salvage master; occurred on weekend when regular port authorities were not available	Ship removed and disposed of; pollution occurred
ESSI ANNE	Freight vessel	Grounded	Pulled off ground by 2 tugs	Channel restrictions made operation time critical	
ESTELLE J./ H9NX	M/Bulk	Aground	Partial offload; refloat		
ETHNOS/ 5MRD	M/V		Refloated with Smit Salvor and Terminales; to Curacao		
EXPLORER	M/T		Dry salvage		

Vessel Name	Vessel Type	Situation	Response	Critical Factors	Outcome
EXXON VALDEZ	T/V	Grounded	80% of cargo saved; lightering, floated from strand		
F/V TENYO MARU		Collision	Several thousand gallons of oil were pumped from vessel after underwater ROV was used to attach hoses	Need for cooperation with Canadian Government; use of ROV	Major oil spill; vessel sank; vessel never salvaged
FAITH 1/ OCEAN 190		Collision	Gasoline cargo in damaged tanks allowed to dissipate before remaining cargo pumped off		
FEDERAL RHINE/ ELBY 3	M/V bulk	Aground	Part offloaded; refloated; SASI	Loaded with grain	
FIJI PRINCESS	Passenger vessel	Aground	Refloated by local salvage company	Rapid response with adequate equipment	
FORT PROVIDENCE /GCPJ	M/V bulk carrier	Aground	Part offloaded; refloated; SASI, anchored		
FRESTON STAR	M/V		Survey		
GARYVILLE	M.T.		Towing		
GEO S	M/V		Dry salvage		
GLACIER BAY	Tanker	Struck uncharted rock	Discharged over the top by equipment from Valdez	Success determined by over the top pumping equipment	Ship returned to service
GUYANA	M/V		Survey		
HAPPY RUNNER	MV heavy lift vessel	Capsized	Refloating operation with Derrick barges		
HAPPY RUNNER	M/V		Survey		
HARP	M/V	Aground	Partial discharge; refloat SMIT/SASI		
HERCULES	M/V		Dry salvage		Scuttled
HERMANOS CARCAMO/ DANT	M/V bulk carrier	Aground	Refloat; LOF		
HYUDAI 12	Bulk carrier	Grounded	Refloated by crew flown in with equipment available in area; ARCTIC SALVOR sailed from Seattle and arrived near end of refloating operation	Personnel skilled in salvage and adaptation of local equipment	Vessel returned to service
HYUNDAI #12	Freight vessel	Grounded	Fuel oil pumped from double bottom tanks and several hundred tons of cargo grain were jettisoned	Unseasonably fair weather; ability to jettison cargo	Refloated
IB906	Barge	Holed double bottom; down-flooded open void hatches while preparing to lock through	Improvised flex hose and drum cofferdam to pump out voids; refloat; "shingle" hole	Complete sinking and hazmat release avoided by running barge into shallow water	Sent to drydock
INDEPEN-DENCE SERVICE	Freight vessel	Grounded; flooded	Barge in tow taken in tow by towing vessels; towed vessel to Palominos Island; capsized while in tow; raised by commercial salvage company		
ISLA PUNA		Grounding			
ITAPUCA/ PPEL	M/V general cargo	Fire in cargo hold #1	Chemical fire extinguished		
JALINGA	M/T		S.T.S.		
JAY GANESH	Freighter	Grounded		Fully loaded; vessel in exposed situation on reef	

Vessel Name	Vessel Type	Situation	Response	Critical Factors	Outcome
JOHN AUGUSTUS ESSBEGER/ DHGJ	M/Bulk	Aground	Refloated SMIT/SASI; LOF		
JUPITER	T/V	Caught fire; exploded; burned	Obtain adequate wire rope to further secure ship; fire extinguished on day 2; reignited and extinguished on day 3; salvage complete on week 5	Obstructed channel in Saginaw River, MI; shorebased firefighting resources could not reach vessel; vessel originally tethered by only one of six original mooring lines; securing ship; extinguishing fire; disposing of contaminated water; alternate uses for available resources	
KEYES 302			Wet salvage		
LADY PAMELA		Sunk	Traditional attempts to raise vessel through pumping out tanks through vents failed due to unique plumbing; crane barge was hired; vessel dragged by barge near shore to take advantage of tidal cycles; after 3 days of massive pumping and lifting the vehicle was salvaged without loss of fuel		
LLOYD GENOVA/ PPJP	M/V general cargo	Aground	LOF; SMIT/SASI		
LOUISIANA BRIMSTONE		Grounding			
M/T ESSO BRUSSELS; S/S SEA WITCH		After collision, ESSO BRUSSELS stern anchor caught in starboard anchor chain of SEA WITCH; SEA WITCH was on fire	SEA WITCH was removed to safe distance from ESSO BRUSSELS; ESSO BRUSSELS fire was extinguished; discharged cargo; continued oil pollution control		
M/V ELDIA	Freighter	Blown ashore	National Strike Force personnel pumped fuel oil from vessel	Cooperation by all parties involved; good weather conditions	Refloated without loss of cargo
M/V SCANDANAVIA SEA	Cruise ship	Fire	Brought to dock to offload passengers; firefighting efforts commenced dockside; vessel almost lost due to excessive use of firefighting water	Lack of personnel trained in shipboard firefighting; time critical need to pump off firefighting water	Vessel saved
M/V STAR CONNECTICUT		Grounded	Navy, Coast Guard, civilian salvors on-site; repaired hull cracks; dewatered pump room	Fully laden vessel; damage to propeller shaft; cracks in double bottom; flooding in pump room; favorable environmental conditions; timely damage control and dewatering operations	Refloated without any oil release
MAASSLOT	M/T		Ship transfer service		
MAASSTAD	M/T		Ship transfer service		
MAASTROOM			Ship transfer service		
MARI BOEING	General cargo	Aground	Total offload; refloat; LOF; SMIT/SASI; towed to Korea		

Vessel Name	Vessel Type	Situation	Response	Critical Factors	Outcome
MARINA L		Grounded	Lightered cargo out with self-unloader vessels and tugs	Fully loaded with linseed oil cake; time critical as building up sand around vessel	
MARINE ELECTRIC/ WOOH	SS bulk	Sank	Diving survey; SASI/Galerne	Cargo coal	
MARQUESA	MV bulkcarrier	Aground	Lightering operation		
MARSHALL KONEV	MV bulkcarrier	Aground	Lightering operation		
MARSHALL KONYEV	M/V	Grounding	Combination of lightering and horsepower freed vessel	Lack of towing vessels quickly available; quickly began silting in; owners decision to only incrementally increase towing vessels; owners were located in Soviet Union; owners representatives were incapable or unwilling to commit themselves to realistic cures for the situation	
MARY LOU	M/V cargo	Aground	Refloated; SMIT/SASI; LOF		
MAVRO VETRANIC	M/V		LOF		
MEGA BORG		Explosion/fire in engine room, pump room and accommodations	LOF		
MELODIC /6ZTY	M/V cargo	Aground	Refloated Smit Cory/McAllister		Docked at Montreal
MERCEDES ENVOY			Dry salvage daily		
MERKUR AMERICA		Engine room fire			
MERMOZ/ FOSK	Passenger ship	Aground	Refloat		
METULA	VLCC	Grounding	Cargo removed by lightering using portable pumps	Weather permitted operation; ship showed remarkable durability in remote, inhospitable part of the world	Pollution limited
MILOS REEFER	Refrigerated cargo carrier	Grounded		Winter weather and damage prevented salvage attempt	Owner removed some oils and hazardous materials
MINERVA	Bulk carrier	Grounded outside channel; unable to refloat; need to lighter	Lightered off cargo to refloat	Difficulty in determining when suitable lightering barge would be available	Refloated
MINORES PRIDE		Aground in mud	Refloated with local salvor and tugs	Fully loaded with iron ore; time critical as blocking river fairway	
MISSISSIPPI QUEEN			Wet salvage		
MONSANTO	Chemical barge	Hit bridge; caught fire			
MONSUN	MV	Collision damage hold #3	Emergency repair		
MOSUN	M/V		Survey		

Vessel Name	Vessel Type	Situation	Response	Critical Factors	Outcome
MOUNT ATHOS / SXLX	M/V general cargo	Aground	Refloated; SASI		
NAFTOPOROS / SYKE	M/V bulk cargo	Aground	Refloated SMIT/SASI; towed to Balboa, Panama		
NO. 6 CHILBOSAN	Refrigerated cargo carrier	Grounded		Winter weather and damage prevented salvage attempt	Owner removed some oils and hazardous materials
NORDIC RIDER / A8UV	Bulk carrier	Aground on reef	Refloat; SMIT/SASI		
NORMAN PACIFIC / S6AL	M/V bulk carrier	Aground	Part offloaded; refloat; SASI	Loaded with grain	
OBELIX / A8WE	M/Tank	Collision	LOF SASI; offloaded; refloated		
OCEAN 255	Barge	Collision with freighter BALSA 37 and barge B 155; spill and fire on B 155	Fire extinguished by Tampa Fire Dept; spill cleanup coordinated by NRC and Maritrans; OCEAN 255 pumped out by Strike Team	Fire	Fire extinguished quickly; good coordination cleanup
OCEAN BEAUTY	Bulk carrier	Grounded	Refloated by SALVAGE CHIEF	Availability of salvage vessel	Returned to service
OCEAN EAGLE	40,000 dwt	Grounding; ship broke in half	USN responsible for bow section, MCS for stern; owner abandoned vessel early so wreck clearance was effected expeditiously; cargo was removed from stern to lightering vessel; cargo remaining in forward half offloaded by pumping into lightering vessel	Pilot not yet aboard	Major pollution occurred; most of oil was salved; both sections were scuttled in deep water by salvors
OCEAN HANNE	Small freight ship	Grounded			
ONDINA	M/T		Dry salvage		
PACIFIC VOYAGER	Fishing vessel	Wreck			Uneconomical to salvage because no reasonable-cost assets available
PENROD HULL			Wet salvage		
PENROD LEGS			Wet salvage		
PERRYVILLE/ KUTN	Product tanker	Aground	LOF SMIT/SASI; refloat; escort to New Orleans		
PETRO SERVICE	T/V	Aground	USCG removed gasoline to small barges; empty vessel dragged off reef	2,000,000 gallons of gasoline on board	
PLATFORM REMOVAL			Wet salvage		
POINT LEVI	Barge	Grounded	Offloaded; refloated; drydocked	Loaded with crude; ice St. Lawrence River	
PRESIDENTE RIVERA		Grounded on rock ledge; holed 7 tanks	Damaged tanks lightered to barges; tugs and engine used to push off ledge; moved to dock; offloaded; temp patches	Crew language barrier; crew incompetence; delays getting lightering barges; access to get salvage pumps on board; 3 knot river currents; tidal rise/fall; good salvage master; loadmaster computer onboard; good vessel condition; high cargo flashpoint	304K gallons of #6 oil spilled; sent to drydock

Vessel Name	Vessel Type	Situation	Response	Critical Factors	Outcome
PRINCE WILLIAM SOUND	VLCC	Lost all power	Local tugs unsuccessfully attempted to take ship under tow; ship's engineers restored power minutes before grounding would have occurred	Fully loaded; drifting toward Columbia Glacier 30nmi away; response effort of local tugs inadequate	No pollution; no damage to ship
PROTESILAUS/GPGA	M/V general cargo	Aground	LOF; refloat; SMIT/SASI		
PSARA	Product carrier	Vessel started taking slow rolls to 15 degrees; came to rest leaning on pier with 8 degree list; lines taught; 2 tugs holding against pier	Careful soundings; calculation of ballast/offload sequence and quantities	Mate used improper offload sequence	
PUERTO RICAN	T/V	Explosion; fire; sinking	USCG ordered vessel to sea; USCG, USN, and local fireboats put out fire; lightering barge; vessel broke in half		Forebody drydocked
RAVEN	Tug	Fell off Marine Railway	Refloated with floating crane	Time critical as blocking ways	
REGENT STAR	Cruise ship	Main switchboard fire; blackout and stranding	Passengers removed by ferry vessel; refloated at high tide with tugs; careful ballast/fuel arrangement	Hull strength due to bending at low tide; empty double bottoms created stability problem for refloating	
RIO TUXPAN/ XCRT	M/V general cargo	Aground	Offload; refloat; LOF		Docked at Vera Cruz
ROBERT E. RESOFF	Fish processor	Broke moorings	Controlled and returned to moorings by fishing vessels	Capable vessels of opportunity willing to help	Returned to service
S.E.W.O.P. MR. DON			Wet salvage		
S.S. SANTA ANNA		Dead in water	Pumped 1,800 gallons fuel onboard to restart auxilliary generators; tug towed ship to Wilmington, NC	Weather; communication; fatigue of crews; equipment failure; lack of specific kind of equipment	Successful rescue
S/S SAN JUAN		Grounding	Channel closed; local tugs assisted; liquid loads were shifted; vessel successfully refloated without pollution	Outstanding salvage master	
S/S SINGAPORE TRADER	Dry cargo ship	High aground	Used Sorbent C Filtration System to filter dirty ballast water in double bottom tanks; "clean water" pumped into St. Lawrence River; SINGAPORE TRADER brought to Montreal		No pollution in Montreal Harbor
SANTA EVON	M/V		Survey		
SARAH FRANK	Oil barge	Sank	Cleanup and pollution control by local vendors; salvage by lifting barge with DonJon Salvage Heavy Lift Floating Crane		Spilled large quantities of oil into Arthur Kill
SEA ALASKA	Fish processor	Dragged anchor and grounded	Refloated by fishing vessels	Success determined by availability of assisting vessels of opportunity	Returned to service
SEA LUCK		Stranded on reefs	No salvor; hired tugs and barges from east coast; lightered cargo to refloat	Fully loaded with phosphates; time critical to refloat before bottom breached on rocks in exposed position	
SEASPAN RIGGER	Log barge	Grounded	Refloated by SALVAGE CHIEF	Broken in half; availability of salvage vessel	Returned to service

Vessel Name	Vessel Type	Situation	Response	Critical Factors	Critical Factors
SEAWITCH STERN	M/V general cargo	Capsized	Drydocked in Brooklyn Navy Yard; complete salvage; towed to Newport News, VA		
SENECA		Suffered severe storm damage; collision bulkhead collapsed; #1CT dumped into forepeak/focsle; propagating cracks in 4 consecutive main butt welds	Examined doublebottoms, focsle, and deck at sea to determine imminent danger; moved vessel to still water in bay; delicate offload over one week	Severe owner interference	Sent ship to drydock
SOVEREIGN OF THE SEAS	Passenger vessel	Caught fire	Crew and local firefighters extinguished fire	Time critical to loss of life	
STAR 6	Tank barge	Intentionally grounded	Owners hired local contractor unable to provide necessary equipment and expertise; USCG lightered cargo; dewatered flooding wing tanks		Vessel moved to drydock
STEELHEAD	Oil production platform	Gas blow-out and fire	Arrival of firefighting tug and two specialized high-output portable pumps (5,000 gpm/ 200 psi) mounted on vessels of opportunity; water used to cool platform structure while relief well was drilled to curtail flow of gas	Fire tug on-site within 18 hours of blow-out; portable pumps arrived within 8 hours of request by owners	"Successful"
STOLT EXCELLENCE	MV tanker	Aground	Lightering operation		
STOLT SINCERITY	Chemical product tanker	Aground	Refloated; SASI		
SUNDANCER	Cruise ship	Sunk after grounding	Refloated by Canadian company	Availability of assets	
SWALLOW	M/V	Grounded	Lightened ship; refloated		
T/B APEX 3417		Collision; vessel sank; closed Houston Ship Canal	Removed from channel with heavy lift cranes after several days;	Economic need to reopen Houston Ship Canal; time delays in completing salvage due to fears of spilling additional oil; financial problems of bankrupt owner; salvage might have been completed sooner had method been available to determine amount of oil remaining in tanks	

Vessel Name	Vessel Type	Situation	Response	Critical Factors	Outcome
T/B STAR 6	Barge	Leaking in its double hull voids and rakes	Tow to port in Texas using small pump to de-water voids; ran out of fuel for pump; barge began to sink; tow decided to return to point of origin (Mobile); heading up mean ship channel barge began to capsize; tow grounded barge which sank in 12 feet of water coming to rest at 45% angle along starboard bilge; local contractor hired to patch holes in, and de-water, voids; equipment and compatibility problems arose due to caustic cargo; Strike Team pumped out voids and rakes; lightered cargo to a second barge	Sunk barge in danger of sliding into and blocking Mobile Ship Canal and cutting off Mobile to all deep draft ship traffic	Sunken barge righted and refloated
TAE WONG	M/V	Aground	Developed method to ignite remaining petroleum products on board and conduct in situ burning response	Limits of liability; age of ship; remoteness of assets	
TAO-191	Navy Fleet Tanker	Dead shift broke adrift from tow; stranded	Refloated by use of tugs and salvor DONJON MARINE	Time critical for vessel in exposed position	
TAURO DEL GOLFO	M/V		Wet salvage		
TENYO MARU	F/V	Sank in 500 feet water	Remotely operated vehicle with attached camera inserted hose into porthole and removed 26,000 gallons of oil		Removed 26,000 gallons of oil
TEXACO NORTH DAKOTA / KFDG	Product tanker	Collision	LOF SMIT/SASI offloaded; refloated	Loaded partly with gasoline	Docked in Port Arthur
TEXISTEPEC	Phosphate bulker	Stranded	Refloated	Time critical to remove from reef before breaking up on rocks in exposed seaway	
TEXISTEPEC /XCTW	M/V general cargo	Aground	Refloated; SMIT/SASI		
THALASSINI EFHI/ELDF6		Aground second time	Refloat; tugs; LOF; SASI		
THEANOULA		Flooded engine room			
TIFOSO	Large tanker	Stranded on reef	Had to get large tugs to site; vessels bottom heavily damaged; CTL; refloated and sunk offshore	No cargo; time critical to refloat before vessel broke up and released bunkers	
TIFOSO/ ABGT	M/Tank	Grounded	Refloated; SMIT, SASI liaison		
TONKAWA	Drilling barge	Capsized	Refloating with derrick barges		
TOPAZ	Small freighter	Stranded on coral	Per USCG, needed to remove bunkers prior to refloat; almost impossible to pump out bunker "C" without heat on dead ship; no salvor, used passing tugs to refloat	No cargo on board; time critical to refloat before rocks breached hull in seaway	

Vessel Name	Vessel Type	Situation	Response	Critical Factors	Outcome
TORREY CANYON	VLCC	Grounding	Salvage Master and RN SupSalv on board; preparations being made to offload and de-water holed tank; explosion occurred incident to salvage preparations; Salvage Master killed; RAF bombed ship against advice of RN SupSalv in unsuccessful attempt to burn remaining cargo	Fully loaded; explosion occurred incident to salvage preparations	Major pollution occurred; ship total loss
TRITON C/ELAP	M/V general cargo	Aground	Part offloaded; SASI, refloated		
TUGS TAMPICO	M/V		Wet salvage		
TULA	M/V		Wet salvage		
USCG MESQITE		Run aground	Successfully picked up after being aground all winter	Lost favorable weather window in December	Picked up in spring, hazardous materials removed, and scuttled
USNS MISSION SAN FRANCISCO	T-2 tanker	Collision; explosion; grounding; channel blockage; stern afloat, bow hard aground, loss of hull modulus in mid-body	Joint response (USN, Merritt Chapman Scott); 102 tons of ballast loaded in stern as stability ballast; operation interrupted due to fog and ice floes; second collision occurred casting stern section loose; 102 tons of stability ballast in addition to mid-body bottom structure was sufficient for adequate stability; protection of environment consisted of state-of-the-art best effort to deploy booms	Sustained severe structural damage	Stern salved; bow and mid-body broken up by explosive cutting and wrecking grabs and disposed of
VEEDOL / ELWR	M/V bulk cargo	Aground	Refloated; SMIT/SASI; LOF	Aground at breakwater	
VENCE-MOS IV		Sank		Full cargo of cement; cement hardened bit by bit; lack of owner cooperation	CTL; wreck removal project
VENTURE LUNA	Cargo vessel	Grounded	Refloated by own efforts; hull holed and spilling oil required owners and P&I to make an OPA 90 response		Continued voyage with temporary repairs
WANDA WHEELOCK	Large ITB barge	Struck rock and sank	Lightened to sister barge; refloated	Obstructing river fairway	
WEEKS #282	Barge	Sank	Refloated with pumps with local salvor	Time critical as blocking river fairway	
WESTERN BARGE			Wet salvage		
WYWURRY	MV bulkcarrier	Aground	Lightering operation		
ZAMIA	M/V		Dry salvage		
ZOE COLLOCO-TRONI	44,000 dwt	Grounding	Offload of cargo using ships pumps, piping, and discharge connection	Weather cooperated; owner supplied lightering vessel which was anchored astern	Environmental pollution was minor

APPENDIX F

CURRENT U.S. NAVY SUPPORT FOR THE COMMERCIAL SALVAGE INDUSTRY

BACKGROUND

Following World War II Congress enacted the Salvage Facilities Act, P. L. 80-513, 10 U.S.C. §§ 7361-7367 "to authorize the Secretary of the Navy to provide salvage facilities, and for other purposes." While there have been no court decisions interpreting this act, the legislative history shows that Congress had several objectives:

- To provide salvage resources to protect the redeployment of government-owned war materiel on chartered ships (but not in excess of national defense needs).
- To foster (but not subsidize) the commercial salvage industry.
- To allow (but not require) the Navy to render salvage services to private vessels when commercial salvors are not available, charging for those services to support the Navy's Salvage facilities.

The Salvage Facilities Act permits the Secretary of the Navy to:

- Provide, by contract or otherwise, salvage facilities for public and private vessels (10 U.S.C. § 7361).
- Acquire or transfer equipment and vessels for operation by private salvage companies (10 U.S.C. § 7362).
- Advance funds to private salvage companies for immediate financing of salvage operations (10 U.S.C. § 7364).
- Settle and receive payment for claims by the Navy for salvage services rendered by the Department of the Navy (10 U.S.C. § 7365).

In the postwar years the Navy provided ships to commercial salvage companies to "support organized offshore salvage facilities" (10 U.S.C. § 7363). This practice continued into the 1970s but was abandoned by 1982 when the Marine Board's report, *Marine Salvage in the United States*, was published.

THE PRESENT SITUATION

The Navy, through the Supervisor of Salvage, Naval Sea Systems Command (SUPSALV), now competitively awards contracts, pursuant to the Salvage Facilities Act, with three commercial salvors to provide salvage facilities for national defense needs around the world:

- Donjon Marine Co., Inc. of Hillside, New Jersey, has responsibility for the Atlantic Ocean, Caribbean, and Mediterranean Seas.

- Crowley Maritime Salvage, Seattle, is the Pacific Ocean contractor.
- Smit Tak, a Netherlands corporation, is the Navy's Western Pacific, Indian Ocean, and Arabian Sea/Arabian Gulf salvor.

The contracts each run for a period of one year with four annual options to renew. The contracts require immediate response with designated facilities and carry a guaranteed minimum value. The contractors receive their costs plus an award fee for operations actually tasked.

SUPSALV also competitively awards contracts with a number of salvage-related companies pursuant to 10 U.S.C. § 7362 to provide services using salvage equipment owned and furnished by the Navy. Military departments are authorized to lease government-owned equipment to private companies under 10 U.S.C. § 2667. Pursuant to this statute the Navy, through SUPSALV, may lease pollution response equipment to private organizations. The terms of these equipment rentals are similar to (but differ in several important respects from) those for salvage equipment leased pursuant to the Salvage Facilities Act.

In each case the contractor is free to enter other contracts with private customers, but may not use the Navy's equipment, and the other contracts must not interfere with the Navy's requirements. These contracts provide for ocean floor search and recovery, in-water ship repair, oil spill response, and salvage equipment depot maintenance; the specific terms vary.

The ocean floor search and recovery contracts are held by Oceaneering Technologies, Inc. This contractor maintains and operates SUPSALV's remotely operated vehicles and underwater search systems. The contracts provide for immediate response and mobilization and pay costs plus award fee for tasked operations.

Oceaneering International, the parent of Oceaneering Technologies, holds the Navy contract to provide worldwide diving services. They include in-water ship inspection and repairs, such as underwater cutting and welding, rudder repair, and propeller and shaft bearing replacement. The contract requires the Navy to furnish certain equipment and requires Oceaneering to provide divers for Navy jobs worldwide. While the contractor is permitted to perform non-Navy work, such work must not interfere with Navy requirements and taskings. The contractor is paid on a daily fee schedule according to the type of diving services provided and recovers its costs plus an award fee.

The salvage equipment depot contractor is Global Phillips Cartner (GPC). GPC maintains the Navy's Emergency Ship Salvage Material (ESSM) warehouses in Williamsburg, Virginia; Stockton, California; Bahrain; Pearl Harbor; Sasebo, Japan; Aberdeen, Scotland; Livorno, Italy; and Singapore. ESSM warehouses hold a large variety and quantity of salvage, diving, and pollution response equipment. The Williamsburg and Stockton facilities also develop and maintain specialized salvage and pollution response equipment. They employ skilled individuals capable of manufacturing, packing, and shipping equipment needed for a salvage and/or pollution response operation anywhere in the world. GPC earns costs plus award fee under this contract. It employs about 130 individuals full time to satisfy contract requirements.

GPC also holds the Navy's pollution response contract and maintains and operates the Navy's extensive inventory of pollution response equipment. On occasion, GPC is tasked by the Navy to support U.S. Coast Guard pollution response operations involving private concerns, shipowners, cargo owners, or pollution liability underwriters. This support almost invariably includes providing Navy-owned pollution response equipment, maintained and operated by GPC, according to a published cost schedule. These costs are billed to the "responsible party," either directly or through the Coast Guard. GPC receives its costs plus an award fee, which are also passed along via the Coast Guard to the customer or "responsible party" for reimbursement.

CONCLUSION

Through the Navy the U.S. government provides limited support for the commercial salvage and salvage-related industries through standing salvage contracts, contracts for salvage-related services, and leases of salvage and pollution abatement equipment.

APPENDIX G

ACCIDENT SCENARIOS

The committee used the accident scenarios in this appendix to analyze the the salvage industry's response in today's environment to similar scenarios in the 1982 study (NRC, 1982). The committee conducted desktop audits and tabletop exercises, and contacted individuals and companies in the industry about equipment and timeliness, which the committee used to assess the salvage posture. The committee wishes to express its thanks to all who participated.

COLLISION AT THE ENTRANCE TO DELAWARE BAY

Technical Issues: Availability of Damage Control, Removal from Strand, Firefighting

While maneuvering to embark a pilot off Cape Henelopen, an inbound, fully laden crude oil tanker is struck within the cargo length by an outbound, laden, break-bulk cargo vessel. The tanker is carrying 122,000 deadweight tons (DWT) of crude oil (Libyan 37° API). The cargo ship is carrying 14,700 DWT of break-bulk cargo, with containers on deck covering the hatches. The situation unfolds at three levels of complexity.

No Fire, No Stranding

There is no fire, although the tanker leaks cargo oil through its breached hull. Both ships retain their propulsion capability, and damage to the cargo vessel is limited to the bow and forecastle. The cargo ship returns to port; the tanker anchors until the oil spillage is suppressed and the cargo is stabilized.

Both Ships Strand

Both ships strand following the collision, and neither ship is able to free itself. The grounding force of the tanker is about 2,000 tons (one foot aground). The tanker is leaking cargo through its breached hull; the cargo ship is not leaking bunkers. Both ships have ballasted down and either have kedged an anchor to seaward or have hooked up a tug astern to prevent broaching. There are no conditions that make this strand particularly hazardous. The bottom is hospitable and it is not a lee shore.

Both Ships Strand, Fire Occurs

Both ships strand following the collision, and fire breaks out on both ships; the spilled oil also is afire. Both ships are stranded beyond their ability to refloat themselves. The cargo ship contains cargo that may be dangerous or may produce a dangerous byproduct when burned.

CHEMICAL BARGE STRANDED OFF PUERTO RICO

Technical Issues: Salvage in Presence of Hazardous Cargo, Underwater Salvage, Refloating

A chemical barge, fully laden with a hazardous polluting substance and inbound to San Juan, Puerto Rico, breaks away from its tow off Isla de Cabras. The barge is double-bottomed, 300 feet long by 43.5 feet, with a depth of 18 feet. It contains 30,000 barrels of cargo in a 5-by-2 tank arrangement. The bottom is hard and the wind is onshore. The situation unfolds at two levels of complexity.

Strand

The barge grounds on a 12-foot pinnacle. The integral cargo tanks are intact. The grounding force is approximately 800 tons; however, the barge cannot be pulled off due to the danger of tearing open the remainder of the bottom. The barge is equipped with pumps, but the generator that drives them is disabled.

Founder

The barge is dragged over the pinnacle, is torn extensively, and sinks in 50 feet of water. The nature of the cargo is such that it must be removed and promptly. There is no leakage of cargo, but all double-bottom tanks and the after-rake compartment are breached. The wing tanks are intact. The barge is upright on a relatively flat, hard bottom with cargo vent valves closed. The underwater visibility is 10 to 20 feet, and all cargo pumping systems, piping, and valves are sound and in operating condition. The pump room is accessible to divers.

LARGE TANKER STRANDED IN THE FLORIDA STRAITS

Technical Issues: Rescue Towing, Refloating, Jettisoning

En route from North Africa to a deepwater port in the Gulf of Mexico, a fully laden 250,000-DWT Very Large Crude Carrier (VLCC) loses power off Sombrero Key in the Florida Straits. She is carrying a full cargo of North African crude oil (44 ° API). The foreign-flag VLCC is owned by a major U.S. oil company. The situation unfolds at two levels of complexity.

Mechanical Breakdown

The VLCC drifts toward shore. If she does not obtain a rescue tow, she ultimately will strand at Delta Shoal. The ship is in a head current of 1.5 to 3 knots, with a 25-knot south-southeast wind. Delta Shoal is 34 miles northeast of the ship when the ship loses power; the overall drift period to Delta Shoal, given the current and wind conditions, is 19 hours.

Stranding

A tug fails to arrive in time, and the ship drifts and strands, with light pollution. The grounding force is 27,000 tons.

CASCO BAY TANKER STRANDING

Technical Issues: Lightering, Rescue Towing, Safe Haven, Damage Stability

Bound for Portland, Maine, from the Caribbean, a 21,000-DWT product tanker carrying a full load of distillate (34° API) strands on a rock off Portland Head Light in Casco Bay, Maine. The tanker is foreign owned and registered. It is fall, and the weather is deteriorating. The situation unfolds at two levels of complexity.

Strand, Refloat

The ship comes off the ground on the incoming tide. The ship is capable of navigation but is leaking. The Coast Guard refuses entry into the port until the ship has been lightered and is not a pollution hazard.

Strand and Refloat, with Damage

The propeller is damaged in the strand such that the vessel is incapable of navigation. The ship is not expected to come off the strand on the incoming tide, which will occur six hours after the initial notification of the stranding. The estimated remaining ground force is 1,350 tons. Major pollution is thought to be from ship's bunkers. Seas are 4 feet, choppy, and building. The weather forecast calls for winds and seas to build for 19 hours, with a northeaster continuing for 48 hours. The ship is leaking from three ruptured center tanks and possibly from the engine room double-bottom bunker tanks.

COLLIER STRANDING OFF NORFOLK

Technical Issues: Refloating, Cargo Offloading, Heavy-lift Capability

Two scenarios are possible. In the first, a number of bulk carriers are anchored in Anchorage L-C to the west of Thimble Shoals Channel waiting to load coal in Norfolk, Virginia. A sudden and fierce summer squall causes one of the colliers to drag its anchor to the east. She is light and strands on rip rap on the eastern bridge island of the Thimble Shoals Channel. The double bottom is flooded to the extent that the ship is about 12,000 tons aground. The ship is projecting into the channel but due to channel width does not block the channel. The second scenario assumes that the bulk carrier is loaded, loses steering, and strands in the same position.

ROLL-ON/ROLL-OFF SHIP CARGO SHIFT

Technical Issues: Rescue Towing, Damage Stability

A roll-on/roll-off (RO/RO) ship carrying heavy vehicles and containers under charter to the Military Sealift Command is caught off Cape Hatteras in heavy weather 40-knot winds from the northeast with 16-foot waves. The ship loses power, is unable to maintain its heading, and takes heavy seas broadside. The heavy rolling causes several vehicles to break loose and eventually puncture the ship. The flooding rate is slow but steady. The weather conditions and the danger of further cargo shifts preclude the crew from taking action to control the flooding. The ship is 20 miles east of Cape Hatteras and 125 miles from the entrance to Chesapeake Bay. If a response cannot be marshalled within 12 hours, the chance of saving the ship is less than 50/50.

GREAT LAKES ORE CARRIER

Technical Issues: Rescue Towing,
Portable Pumping Capability, Damage Stability

In the late fall, a Great Lakes bulk carrier loaded with 25,000 tons of iron ore is caught in a severe storm in Lake Superior, near the U.S./Canada border. She suffers severe structural damage, such that the longitudinal strength of the ship is jeopardized. Weather is overcast in rain, winds are 35 to 50 knots, and waves are 12 to 15 feet. The ship is experiencing flooding and has no motive power.

COLLISION OF TANKER AND BULK GRAIN CARRIER OFF GALVESTON

Technical Issues: Firefighting (with Inerting), Rescue Tow, Lightering

A bulk carrier (U.S. flag and cargo) hits a foreign-flag tanker (Far East owner, U.S. cargo) 2 miles off Galveston, Texas. The bulk carrier's bow penetrates the tanker in the midship tank. Fires on both ships cause considerable structural damage. The tanker pump room, emergency power, and piping are intact. The 150,000 DWT tanker is loaded with North African crude; the 70,000 DWT bulk carrier is loaded with grain for Europe. The tanker has neither an inert gas system nor a ship's drawing or stability manual. Fires prevent access to both ships' bows for anchoring. Crude oil is leaking and partially burning from the No. 1 starboard wing tank and the No. 3 center tank.

LOSS OF PROPULSION ON VLCC IN OPEN SEAS

Technical Issues: Rescue Towing

A 300,000-DWT VLCC (U.S. flag and cargo) loses propulsion while inbound from the Caribbean through the Yucatan Straits to the Louisiana Offshore Oil Port (LOOP). An approaching hurricane threatens to drive the ship toward the northwest shore of the Yucatan Peninsula or into the Bay of Campeche. Only the main propulsion engines are affected; emergency power and other facilities are operable. The casualty site is about 190 miles south-southeast of LOOP, 425 miles north-northeast of Merida, and 600 miles west-northwest of Key West. The hurricane is 575 miles southeast of the VLCC, moving north-northwest at 15 knots, and can veer east or west.

Winds are at 40 knots from the northeast, waves at 16 feet from the northeast, the current is easterly at 2 to 4 knots, and the forecast is uncertain. Even with major veering of the hurricane, the VLCC can be expected to hit Yucatan in about 70 hours. Although the ship has power, its anchors are useless because the water is over 6,000-feet deep.

TANKER RAMMING OF UNMANNED PLATFORM

Technical Issues: Firefighting, Rescue Towing, Rig Salvage

A product tanker (U.S. flag and cargo) rams an unmanned six-well oil and gas production platform 70 miles south of Morgan City, Louisiana. The tanker bow is badly damaged and an empty wing tank is ruptured, but the rest of the cargo space is intact. Fire breaks out due to oil and gas escaping from production equipment and pipeline. The ship loses steering and the forepeak is inaccessible due to fire and inoperable due to damage. The platform emergency shutdown system fails on at least one well, so oil continues to feed the fire. The tanker is unable to separate from the platform for fear of increasing the damage.

COLLISION OF BENZENE TANKER AND CONTAINER SHIP WITH NUCLEAR MATERIAL ABOARD

Technical Issues: Salvage in Presence of Hazardous Materials, Safe Haven, Cargo Offloading, Firefighting

A "drug-store" tanker (U.S. flag and cargo) is struck by a U.S.-flag container ship that has, among other cargo, one 50-ton container of spent nuclear fuel. The tanker is holed in a benzene cargo tank and fire breaks out on both ships. The casualty occurs in the open sea, 200 miles south of Mobile, Alabama. Both ships' bows are inaccessible due to fire; the container ship forepeak machinery and windlasses are inoperable. The container ship is flooding.

LOSS OF STEERING ON AMMONIA TANKER, GROUNDED OFF APALACHICOLA

Technical Issues: Salvage in Presence of Hazardous Materials, Refloating, Lightering, Rescue Tow

A 30,000-DWT anhydrous ammonia tanker has a steering failure off the coast between Tampa and Panama City, Florida. She is driven aground in shallow waters off Apalachicola, bow first at midtide. The double bottom is ruptured forward; cargo tanks are intact but subject to increased heat transfer from water in the hold and potential damage from motion induced by seas. Condition of onboard protective clothing and breathing gear is questionable. The relief valves on the tanks are set at 10 pounds per square inch. The ground reaction on the strand is approximately 1,500 tons at high tide; the ship must be lightened by about 3,000 tons to break her loose. The wind is 20 knots with waves to 6 feet, and the forecast is for worsening weather conditions.

TANKER RAMMING OF JACK-UP MOBILE DRILLING RIG NEAR SHIPPING FAIRWAY SOUTH OF LAKE CHARLES

Technical Issues: Firefighting, Inerting Equipment, Lightering, Multiple Casualty Salvage

A 200,000-DWT tanker strays from the shipping fairway on a foggy night and rams a three-leg, jack-up mobile drilling rig with 400-foot water depth capability. Due to the summer hurricane season, the jack-up deck is set to have 51-foot clearance above the sea level. The tanker's bow goes beneath the jack-up deck and damages the legs. Primary tanker damage is to the forecastle deck and machinery and to the forepeak and bulbous bow. The forepeak tank is ruptured but all cargo tank piping is intact. The drill rig damage is major, although the legs have some residual strength. The drill string is ruptured, but high enough above the seafloor blowout preventer that oil flow can be stopped by diver operation of the blowout preventer. There is no fire, because no free oil was present before metal-to-metal motion was stopped. There is the potential of a gas blowout, and spilled oil is polluting nearby fishing grounds.

TANKER BREAKS LOOSE FROM MOORING

Technical Issues: Rescue Tow, Lightering, Damage Stability, Jettisoning

A 100,000-DWT crude tanker breaks loose from its mooring off Oahu, Hawaii, owing to failure of mooring attachments. Winds are 25 knots and there is heavy rain. Derangement of the propulsion plant prevents lighting off the main engine; emergency power is available initially. The ship strands in a broached condition and is moving slightly on the strand. The ship is 12,000 tons aground and will go further aground within 36 hours, suffering major structural damage. Tugs are unable to pull the ship from the strand. Weather is changing to favorable conditions.

RO/RO SHIP RAMS PLATFORM IN COOK INLET

Technical Issues: Rescue Tow, Heavy Lift,
Internal Transfer of Ballast, Salvage in Ice

A RO/RO general cargo ship inbound from Seattle to Anchorage is caught and stopped by ice flowing at the top of the tide in Cook Inlet, Alaska, and is swept into an oil platform at night. The ship strikes the platform beam, and the platform's several projections puncture the ship's side. The engine room and machinery spaces are flooded and the ship is expected to heel severely in three to four hours. There is no pollution threat. The ship is swept free of the platform but is unable to maneuver.

TANKER AND CONTAINER SHIP COLLIDE OFF STRAIT OF JUAN DE FUCA

Technical Issues: Rescue Towing, Lightering,
Damage Stability, Multiple Response, Safe Haven

Two inbound ships, a 35,000-DWT container ship bound for Vancouver, British Columbia, and a 125,000-DWT crude tanker inbound for Puget Sound, collide 50 miles off the entrance to the Strait of Juan de Fuca. An engine room fire aboard the tanker has distracted the crew, contributing to the accident, and the fire rapidly flames out of control with resulting loss of propulsion. The crew abandons the ship. The container ship, badly damaged, can maneuver but may need rescue towing to make it into port. The damage to the tanker is confined mostly forward of the collision bulkhead. The fire in the engine room burns out. The tanker has been refused entry into a port until the first cargo hold aft of the collision bulkhead is removed at sea.

TANKER AND FISH PROCESSOR COLLIDE IN SHELIKOF STRAIT

Technical Issues: Salvage in Presence of Hazardous Materials,
Ship to Ship Transfer, Safe Havens

A 930-foot liquified natural gas (LNG) tanker with 127,000 cubic meters of cargo, outbound from Nikiski, Alaska, to Japan, rams a 600- to 700-foot Korean-flag fish processor on station off Kodiak Island, Alaska. The fish processor is holed in two cargo holds; the ammonia system is damaged and leaking, forcing all hands on deck for evacuation. The captain beaches the fish processor deliberately in Wide Bay.

The LNG ship is holed in the bow, but no cargo tanks are leaking. The captain is unwilling to proceed with the voyage, and the vessel will have to transfer its cargo to another LNG tanker. The transfer should take place in protected waters, but the state of Alaska may be reluctant to allow the ship to return.

STRUCTURAL FAILURE ON ANHYDROUS AMMONIA CARRIER IN THE COLUMBIA RIVER

Technical Issues: Rescue Towing, Salvage in Presence of Hazardous Cargo, Refloating

A ship carrying anhydrous ammonia, 630-feet long with a 90-foot beam and 31-foot draft, has a structural failure and beaches on the Columbia River bar. Spring flood tides generating a 9-knot current and heavy seas combine to cause waves to break over the bar, and the ship is damaged. The potential for an ammonia cloud poses a threat to Astoria, Oregon. The main switchboard of the vessel has been shorted out, causing the casualty.

STRUCTURAL DAMAGE TO A PRODUCT TANKER IN SAN FRANCISCO BAY

Technical Issues: Rescue Towing, Safe Haven, Salvage in Presence of Hazardous Cargo, Underwater Salvage

A 28,000-DWT chemical tanker loaded with benzene and bound from San Francisco to the East Coast is overstressed due to improper loading. Passing over the bar at the harbor entrance, a crack opens athwartship on the main deck. The master attempts to turn around at the channel's edge, the crack proliferates, and the tanker breaks in two. About 1,000 tons of benzene are released into the water. The stern section is unstable and sinks in 40 feet of water, while the bow section, containing some 10,000 tons of benzene in four tanks, remains adrift.

FIRE AND EXPLOSION ON CHEMICAL TANKER OFF LONG BEACH

Technical Issues: Firefighting, Damage Stability, Lightering, Heavy Lift

A 35,000-DWT product tanker grounds on the breakwater at the entrance to Long Beach, California, catches fire, and explodes on a smoggy summer day. At least one or more of the forward tanks has been ruptured. The ship is fast to the breakwater.

FIRE ON A TANKER OFF VALDEZ

Technical Issues: Rescue Towing, Firefighting

A 225,000-DWT tanker departs Valdez, Alaska, assisted by three 7,000-horsepower tugs. The tugs are equipped with gear for rescue towing and firefighting. Fire breaks out in the engine room, fed by a broken lube oil line, forcing evacuation of the engine and control rooms and the release of carbon dioxide into the engine room. The escort tug has left to return to port. The ship begins to lose steering and swing into the trough. The use of the ship's anchors is attempted unsuccessfully. The fire is confined to the engine room through inerting of the pump room and cofferdams.

TANKER STRANDING OFF EAST COAST[1]

Technical Issues: Jettisoning, Damage Stability, Refloating

An 80,000-DWT tanker is en route from Pajaritos, Mexico, to a refinery on the Delaware River with a cargo of 78,500 tons of Isthmus crude oil. The ship experiences a mechanical problem with its gyro compass and grounds in 35 feet of water near Diamond Shoal. The ship runs head-on into the shoal at 15 knots and comes to rest with 45 percent of her length resting on the sand. The hull remains intact and there are no leaks. A hurricane is approaching and will arrive within 16 hours. The increasing force of the winds and waves is expected to force the ship further onto the shoal, with eventual structural failure due to very high bending stresses. Internal transfer of the cargo will not refloat the vessel, and lightering is not possible due the deteriorating weather and distance from lightering resources. The discharge of 1,700 tons of cargo will allow the ship to refloat and continue its voyage. Delay will result in the ship being damaged beyond the point where she and her cargo can be saved.

[1] This scenario was not used in the 1982 study and is included here to draw attention to the issue of jettisoning. While the incident unfolds on the East Coast, the scenario could apply to any coast or region.

APPENDIX H

U.S. LAWS AFFECTING SALVAGE

AUTHORITY	PURPOSE	COMMENT
Salvage Facilities Furnished by Navy P.L. 513, 4 May 1948 10 USC 7361-7367	The act authorizes the Secretary of the Navy to provide salvage facilities by contract or otherwise, to transfer or charter salvage vessels and equipment for operation by private salvage companies, to advance funds to private salvage companies, to finance salvage operations, and to collect fees for salvage services.	Having given the Secretary of the Navy authority to act in the salvage field, it may be inferred that Congress expected him to use that authority. It has been so used over the years in support of commercial salvage activities. However, it may not be inferred that the Salvage Act requires the Secretary of the Navy to act. Nor does it require the maintenance of a certain level of salvage capability or type of posture. The statutory authority does not obligate the Navy to maintain salvage facilities in excess of its own needs or to render assistance on all occasions. Such a position may be necessary to avoid open-ended liability and exposure to claims from shipowners who were not salved. (See 32 CFR 754.2 (g).) The Act was passed shortly after World War II. The Navy Department was designated as the logical salvage agency for the following reasons: • The Navy already had a (military) salvage organization and therefore understood all phases of the problem. • It was deemed necessary to continue Navy interest in salvage in times of peace so that sufficient personnel and equipment would be available in the event of war. • The Navy had sufficient public salvage vessels to cover waters where private salvage enterprises were not prone to operate. • The Navy was the primary source of mariners and engineers trained in salvage operations.

AUTHORITY	PURPOSE	COMMENT
Saving Life and Property Act of 4 August 1949 14 USC 88	Complete revision of Coast Guard authorities. Authorizes the Coast Guard, in the broadest possible terms without limitation as to method or place, to save lives and property.	The language, "to perform any and all acts necessary to rescue and aid persons and protect and save property" is broad enough to encompass salvage questions. However, the Coast Guard has consistently maintained that it is not in the salvage business. The Coast Guard does not have any substantial salvage capability in the sense of conducting a major offshore salvage operation. The Coast Guard's emergency response capability is largely in the area of search and rescue (saving of life) and marine environmental protection (oil spill response). Related to these are their major activities in safety of navigation, and responsibilities under the National Contingency Plan for overseeing marine emergency response.
Clean Water Act (33 USC 1251 et. seq.), including Sec. 1321 "Oil and Hazardous Substance Liability"	Establishes U.S. policy that there shall be no discharges of oil or hazardous substances in waters under U.S. jurisdiction (including the Fisheries Conservation Zone (200 miles)), and authorizes executive actions to that end.	Authority extends throughout 200-mile zone. Provides basic operating authority for Coast Guard's marine environmental protection activities and for the National Contingency Plan. President authorized to direct all public and private efforts to prevent marine pollution whenever a marine disaster has created a substantial threat of a pollution hazard. Efforts may include removal or destruction of the vessel posing the threat. President authorized to clean up spills, unless, pursuant to the National Contingency Plan, the owner or operator is taking proper action. The difference between the two authorities noted above is that the President does not have to take account of the owner or operator's actions if there has been a maritime disaster.
Intervention on the High Seas Act 5 February 1974 33 USC 1471-1487	Incorporates into U.S. law the international convention relating to intervention on the high seas in case of oil pollution casualties.	Authorizes the Coast Guard to take whatever measures are necessary to prevent or eliminate danger to the coastline of the United States from pollution from a marine casualty on the high seas. Range of possible actions includes removal or destruction of the ship or cargo that is the source of the danger.

AUTHORITY	PURPOSE	COMMENT
33 USC 409, 414, 415, Wreck Statute	These three statutes provide the U.S. Army Corps of Engineers with authority to remove wrecks from the navigable waters of the United States. • 33 USC 409 makes it unlawful to obstruct navigable channels. It sets forth the duty of the owner of a sunken craft to mark and remove it. Failure to do so is considered an abandonment of such craft, subjecting it to removal by the United States. • 33 USC 414 provides for removal by the Secretary of the Army of sunken wreck obstructing navigation. It contains provisions for notice to the owner and authorizes the Secretary of the Army to contract for removal. • 33 USC 415 provides for summary removal in emergency cases. When an obstructing vessel or craft seriously interferes with or especially endangers navigation, the Secretary of the Army may take immediate possession of such craft and remove or destroy it and clear the waters of the obstruction.	The Corps of Engineers, U.S. Army, has been successfully operating under the Wreck Statute for more than 80 years in clearing the navigable waters of the United States of obstructions caused by sunken vessels. Removal of wrecks by the Corps of Engineers is generally confined to those considered obstructive to general navigation. The Corps of Engineers has its own contracting authority and some in-house capability for wreck removal. In time-critical situations, the Corps may request Navy assistance and/or obtain assistance under an existing Navy salvage contract.
P.L. 96-387 (94 Stat. 1545) National Defense Features	Authorizes the Secretary of Commerce to equip certain U.S. vessels with national defense features.[1]	A possible federal assistance mechanism for improving the salvageability of U.S. ships.
Merchant Marine Act of 1936 46 USC 1192 (Construction, Reconstruction, Remodeling)	Authorizes the Maritime Administration to construct, recondition, or remodel vessels in private or public shipyards.	A possible mechanism for strengthening the U.S. salvage fleet.
46 USC 1273 (Obligations, Guaranteed Payment)	Authorizes the Secretary of Commerce to guarantee private financing of vessels.[1]	A possible mechanism for strengthening the U.S. salvage fleet.
"Cabotage Law" (Act of 11 June 1940) 46 USC 316	Restricts the activities of foreign tugs and salvage vessels in U.S. navigable waters.	Approval of a high customs official is required in order for foreign salvage vessels to work in coastal waters of the United States. The effect of the law is to make it very difficult to utilize foreign salvage vessels on a timely basis, even though such assistance may be the only kind available in an emergency. Coverage may not apply to Alaska or Hawaii.
46 USC 725 Canadian Vessels Aiding Vessels Wrecked or Disabled in U.S. Waters	Authorizes Canadian vessels to render aid or assistance to Canadian or other vessels wrecked or disabled in the waters of the United States contiguous to Canada, and vice versa.	This statute, together with a 1908 treaty, allows Canadian salvors to operate in waters of the United States contiguous to Canada, in return for reciprocal privileges for United States salvors. The Cabotage Law exempts salvage operations authorized by treaty or by 46 USC 725.
Salvage Act of 1912 46 USC 727-731	To harmonize U.S. law with the Salvage Convention of 1910 (Brussels Convention). The convention establishes arrangements governing the conduct of salvage operations and the obtaining of compensation for them.	See the Salvage Convention of 1989.

AUTHORITY	PURPOSE	COMMENT
Suits in Admiralty Act 46 USC 741-752	The intent of the act is to subject the United States to the same liabilities, apart from seizure, as are imposed by law on the private shipowner. Sec. 10 of the act authorizes the United States and the crew of any merchant vessel owned or operated by the United States to sue for compensation for salvage services rendered by such vessel and crew.	This act, together with Public Vessels Act, constitutes a broad, consistent, and complete waiver of the government's sovereign immunity, with the exception that public vessels cannot be seized or attacked through court action. This exemption from seizure is relevant to salvage because salvage awards are determined after the fact. However, it can be argued that it is unnecessary to seize a U.S. vessel in order to achieve payment on a salvage award.
Public Vessels Act 46 USC 781-790	Grants a right of action for damages caused by government vessels of the United States. Affords claimants a legal remedy for damages by public vessels including salvage claims against public vessels.	The Public Vessels Act deals only with suits against the United States, including actions for compensation for towing and salvage services rendered to public vessels. This act has nothing to do with affirmative Navy claims for salvage services rendered by the Navy, which are the usual situations where Navy salvage forces are involved. In a recent court case (*Julius A. Furer* litigation), it was held that suits against the United States must conform strictly to this statute. It follows that a commanding officer lacks the authority to sign a Lloyd's open form salvage agreement, which would commit the United States to arbitration in London and give the salvor a maritime lien.
Act of 3 July 1944, amended by Act of 10 August 1956 10 USC 7721-7730 10 USC 7622	Provides the Navy with authority to stay judicial proceedings under the Public Vessels Act in time of war. Also provides the Navy with authority to settle claims for damages caused by Naval vessels without litigation. These include claims for compensation for towage and salvage service, including contract salvage, rendered to a vessel in the naval service or to other property under the jurisdiction of the Navy.	Enables claimants against the Navy to settle admiralty claims without having to resort to litigation. However, should satisfactory settlement prove to be impossible, the necessary authority to sue is provided by the Public Vessels Act.
33 USC 1221-1227 Ports and Waterways Safety Act of 1972	Authorizes the U.S. Coast Guard to establish Vessel Traffic Services and Systems for ports, harbors, and other congested waters.	The Coast Guard may require that vessels use or otherwise comply with established port and waterway safety procedures. Special procedures, such as restricting vessel operations, may be applied in particular circumstances, such as in the movement of hazardous cargoes or in adverse environmental conditions.

AUTHORITY	PURPOSE	COMMENT
P.L. 96-510 Comprehensive Environmental Response, Compensation, and Liability Act of 1980	Provides for response to hazardous substances emergencies.	Establishes a tax on hazardous substances and a Hazardous Response Trust Fund. Provides that the National Contingency Plan authorized under Sec. 311(c)(2) of the Federal Water Pollution Control Act of 1972 shall include a National Hazardous Substances Response Plan. Requires that the National Response Center be notified of all unauthorized releases of hazardous substances. Authorizes the President to act, consistent with the National Contingency Plan, to respond to hazardous substances emergencies unless the President determines that a responsible party is responding satisfactorily. Authorizes the President to initiate abatement actions in response to the threat of a hazardous substance emergency; assigns liabilities and establishes rules of financial responsibility.
Recaptures: Award of Salvage Costs, and Expenses 10 USC 7672	Applies in the event that the Navy recaptures, before condemnation as a prize, a vessel that has been seized by the enemy. States the duty of the Court, the disposition to be made of the recaptured property, and provides that the amounts awarded as salvage shall be paid to the United States.	When a vessel is captured by the enemy in time of war, the question of her seizure and subsequent condemnation or release is for the courts of the captor. The purpose of bringing in a captured ship or cargo for adjudication is to have a sentence of condemnation pronounced by a proper tribunal, a Prize Court, declaring the capture to have been made properly. Such a decree is necessary to vest the property in the captor. The proceedings are *in rem* and they transfer a title to the property that should be universally recognized, if the Prize Court has jurisdiction. However, if the ship is recaptured before condemnation as prize, she has no such status. The situation then is of a captured ship that has been recaptured, with no change in her original status. Recapture of a vessel or property from an enemy, pirate, or privateer has long been recognized as a salvage service and the subject of a salvage award. Since possession of the owner was displaced by the capture, restoration of the property to him or her is a beneficial service, resulting in a salvage award.
Seamen's Suits 28 USC 1916	Relieves seamen from prepayment of court fees, costs, or security in suits that they bring concerning wages or salvage or the enforcement of health and safety laws enacted for their benefit.	This statute looks back to former times when seamen were believed to be impecunious, improvident, and imposed upon by their employers. As such, they were considered wards of the admiralty.

AUTHORITY	PURPOSE	COMMENT
Tariff Act of 1930 Provisions Regarding Cargo of Wrecked Vessels 19 USC 1483 19 USC 1310	19 USC 1310 allows free importation of merchandise recovered from sunken and abandoned vessels. 19 USC 1483 allows the wrecked merchandise to be entered and cleared through customs, leaving the rights of ownership and other claims to be determined by court order or other proceedings.	The procedures for entering and clearing cargo from a wrecked vessel are as follows: • The vessel must have been sunk in U.S. waters for two years or more. • The vessel must have been abandoned by the owner. • The salvor must raise such vessel (or, presumably, retrieve the cargo). • The salvor must enter the merchandise in the applicable customs district. The salvor of the cargo is regarded as the consignee for customs purposes.
Wrecked Vessels Act of 24 February 1915 46 USC 14	Authorizes the U.S. registration of foreign-built vessels that have been wrecked in U.S. waters; purchased by U.S. citizens (such as U.S. salvors); and repaired in U.S. shipyards (at the expense of U.S. owners), so long as the value of repairs equals three times the appraised salved value of the vessel.	It is not believed that vessels qualifying for documentation under these provisions represent a significant addition to the U.S. Merchant Marine.
Agreement as to Loss of Lien or Right to Wages 46 USC 600	Protects the seamen's lien for wages. Provides that any stipulation by which a master or seaman consents to abandon his or her right to wages, or to abandon any right to salvage, is wholly inoperative.	Crew members of ships regularly engaged in salvage are not entitled to salvage awards. This is because the crew members perform work that they may have been hired to do. As such, they are not volunteers, and voluntarism has traditionally been an essential element for a salvor to qualify for a salvage award. Since this element of voluntarism is absent in the case of professional salvage crews, they do not have any rights in the nature of salvage, and thus are not within the purview of the statute.
Attachment or Arrestment of Wages 46 USC 601	Protect seamen's wages and salvage awards from seizure.	This statute protects the sailor against signing away his or her potential future salvage rights.
Plunder of Distressed Vessel 18 USC 1658	Prohibits and punishes plundering a distressed or wrecked vessel within the jurisdiction of the United States; obstructing the escape of any person trying to save his or her life from such a vessel; or rigging false lights or extinguishing true lights with intent to bring a vessel into danger.	This is the only section of the U.S. Criminal Code directly related to salvage. It is in the chapter dealing with piracy and privateering.
46 USC 721 Vessels Stranded on Foreign Coasts	Provides that the U.S. government, through its consuls, will assist U.S. vessels that are in extremis on foreign shores for the purpose of saving the vessel, its cargo, and other effects and delivering them to the owners. When the master, owner, or consignee is present or otherwise capable of taking possession, the consul shall not exercise such authority.	The settlement of salvage liens takes precedence over such consular actions.

AUTHORITY	PURPOSE	COMMENT
46 USC 722-724 Wrecks on the Coast of Florida	Regulates the disposition of salved property on the coast of Florida by licensing vessels and masters regularly employed in wrecking.	These statutes hark back to the days when there was considerable salvage and wrecking activity on the Florida Keys and coasts. These sections were lucrative business for American salvors. The U.S. cabotage laws, especially 46 USC 316(d), treat similar matters in a more broadly applicable way.
Oil Pollution Act of 1990 (P.L. 101-380, 104 Stat. 484 (1990))	To amend the Clean Water Act to provide increased and additional liability for oil pollution of the marine environment.	(Only provisions of most interest in the context of salvage will be commented on here.) Unlike the Clean Water Act, the President is charged with ensuring effective and immediate response to a threatened or actual spill. The effect of this amendment is to make all spills "federal" without regard to the actions being taken by the owner or operator of the vessel. Thus a federal on-scene coordinator will coordinate spill (or threatened spill) response from the outset. Contains an exemption from liability for actions taken or omitted in the course of rendering care, assistance, or advice consistent with the National Contingency Plan. Provides for a substantial increase in civil penalties and criminal sanctions. Expressly does not preempt states from imposing additional liability or requirements. Provides that limitation of liability under the Limitation of Liability Act, 46 USC App. sec. 183-196 does not affect liability, fines or penalties imposed by OPA 90. This has been interpreted to mean that Limitation Act protections and procedures are not available in spills to which OPA 90 applies. The salvor may be strictly liable as a responsible party in the first instance (e.g., if he is held to be the legal "operator" of the casualty or if he spills oil from a lighter that he owns or operates in the course of a response) or as a "third party" responsible party under the OPA 90 defenses to liability.
Act to Prevent Pollution from Ships, 33 USC secs. 1901-1911 (implementing the Protocol of 1978 Relating to the International Convention for the Prevention of Pollution from Ships, 1973 (MARPOL 73/78)	To implement an international convention prohibiting marine pollution from oceangoing vessels.	Permits jettison of oil or oily mixture necessary to save a ship or life at sea. This provision was limited under the pre-OPA 90 Clean Water Act (Federal Water Pollution Control Act) to the Exclusive Economic Zone (EEZ). OPA 90 amendments to the Clean Water Act further limit its utility by making discharges in the EEZ subject to cleanup costs and damages (but not fines and penalties). Thus jettison remains a limited option under MARPOL 73/78.
International Convention on Salvage (1989) Done at London, April 28, 1989	To modernize the international law of salvage, particularly to provide specifically for compensation for salvors' efforts to prevent or mitigate damage to the environment.	The treaty has been ratified by the United States, but as of 16 December 1993 had not come into force, lacking the requisite 15 accessions or ratifications. The treaty will be self-executing when it enters into force.

AUTHORITY	PURPOSE	COMMENT
International Convention on Salvage (continued)		To date the only enabling legislation is 105 Stat. 2225, sec. 40 (codified at 46 USC sec. 729, and 46 USC sec. 731.) 46 USC App. sec 729 enacts the provisions of Article 16 (2) of the Convention pertaining to compensation for life salvage. 46 USC sec. 731 makes prior provisions regarding the right to compensation for salvage when vessels are owned in common (46 USC sec. 727); life salvage (46 USC App. sec. 729); limitation of actions (46 USC App. sec. 730); and duty to render assistance (46 USC sec. 2304) inapplicable to warships and "government ships appropriated exclusively to a public service." The special compensation provided in Article 14 of the convention and the criterion of "the skill and efforts of the salvors in preventing or minimizing damage to the environment" in fixing a salvage reward set forth in Article 13 (1) (b) of the convention had not been incorporated into U.S. statutory law as of 16 December 1993.

1. Authority transferred to Secretary of Transportation in 1982.

APPENDIX I

SALVAGE TUG REQUIREMENTS

GENERAL

The power required by a tug to tow a ship is a function of many variables. First, the size of the ship and its general characteristics influence the effect of the wind and wave forces acting upon it. Second, the size and characteristics of the tug as well as the type of propulsion are significant factors in determining not only the magnitude of the environmental forces acting on it but also its ability to operate in that environment. Last, but not least, the towline arrangement and the experience and expertise of the tug master and crew have a large bearing on the ability to tow in heavy weather.

When evaluating the ability of a tug to tow a vessel it is necessary to understand the dynamics of the situation. A tug towing a tanker can be represented for analysis purposes by three components:

1. A mass representing the ship, which is acted upon by time varying accelerations, damping, and spring coefficients.
2. A nonlinear spring representing the towline.
3. A mass representing the tug, which is also acted upon by time varying accelerations, damping, and spring coefficients.

This is a very complex system both from a theoretical point of view as well as in actual operation. The multiplicity of time varying forces and resulting motions combined with the nonlinearity of the towline can lead to extremely high dynamic loads on the system in heavy seas.

While one might prefer to think that any size vessel could be towed in any weather with a large enough tug, this is simply not the case. The limiting factor is a function of the relative motions between the two vessels and the ability of the towline and towing winch to allow for that relative motion without breaking, while still providing sufficient towing force to keep the vessel moving or at least to maintain its position. It was demonstrated—most dramatically in the case of the *Amoco Cadiz*—that even a world-class tug cannot tow a vessel under some conditions. The relative motions between the two vessels simply becomes too great for the towline's ability to stretch. Attaching a towline is a difficult and dangerous operation in storm conditions and may not be possible, again, due to relative motion between the vessels.

Tugs use several different methods to provide the stretch required in a towing system. First, an ocean towing winch may be designed to automatically pay out and heave line as the tension varies. This helps to reduce major shock loads. Next, the length and weight of the tow wire results in a catenary, which allows for some degree of relative motion between the tug and ship. Finally, the use of a doubled nylon hawser as a shock absorber within the towline adds stretch to the system.[1,2] Together these allow substantial relative motion between the two vessels. While other methods are used to allow for relative motion between the vessels, the above are the most common.

used to allow for relative motion between the vessels, the above are the most common. However, as the relative motion nears the limits of the winch, catenary, and hawser system capabilities, the towing arrangement will either become very stiff, with consequent large shock loads, or the winch will pay out more line than it can recover. In either case, the tug will be unable to continue towing.

Salvage Scenario

To estimate required tug horsepower, it is assumed that a vessel has lost all power upwind of a lee shore, and that the tug would be required to tow the vessel head into the wind and waves or at least maintain position relative to the lee shore.

Analysis Methodology

The calculations used in this study are based on published information relating to the behavior of tank vessels in various wind and sea conditions. A relatively simple procedure was used to analyze the size of tug required to tow various size tankers. This consisted of calculating the mean wind and wave forces on the tanker and tug and converting them into required bollard pull. This is a static analysis, not a dynamic analysis, and as such assumes that the towing system consisting of the towing winch, towline, and spring make adequate provision for relative motion between the two vessels. The calculation of the wind and wave forces is described below.

Wind and Wave Forces

In analyzing the forces on a large ship in storm conditions, it is necessary to estimate the actual sea state in terms of significant wave height and period, because wave drift forces acting on a vessel increase as the wave height increases and usually decrease as the wave period increases. Figure I-1[3] provides data on sea states relative to wind speed for fully developed seas. A fully developed sea is one in which the wind has been blowing long enough and far enough for the waves to develop to their maximum extent. Footnote a to this table notes that this rarely occurs for winds in excess of 50 knots. As shown in the table, a minimum duration of 69 hours and a fetch of 1,420 nautical miles would be required for a sea to become fully developed in 50-knot winds.

For this study, wave height and period statistics were taken from *Global Wave Statistics* by British Maritime Technology Limited.[4] This book provides wave height and period probability distributions worldwide. These probability distributions are tied to specific areas of the world and specific wind directions. These data allow us to calculate the joint probability of various wave heights occurring at the same time as onshore winds for different regions of the United States. Graphs of the calculated joint probability are attached (Figures I-2 through I-6).

Based on a range of typical sea states, the mean wave drift forces were estimated using OCMOTA,[5] ship motion and wave force prediction program developed by the Maritime Research Institute of the Netherlands (MARIN) for the Oil Companies International Marine Forum (OCIMF). To ensure that the probable range of wave heights and periods were fully covered, two different average wave periods were used for each wave height. This was done since wave drift forces are sensitive to wave period. The curves of wave drift force presented (Figure I-7) represent the mean of these two conditions. It should be noted that the wave energy spectrum used was a Jonswap spectrum with a peak enhancement factor of 3.3. This is representative of a building storm. The wave drift forces on a ship in a building storm are normally much higher than in a fully developed sea of the same height because of the shorter mean wave periods. Wind forces were calculated based on *Prediction of Wind and*

Current Loads on VLCC's by OCIMF[6] (Figure I-8). The sum of these forces together with the forces acting on the tug gives a reasonable estimate of the minimum bollard pull required to control a tanker in storm conditions.

Estimation of Required Tug Power

After calculating the wind and wave loads acting on the tankers, the tug/towwire resistance was added and an estimated loss of efficiency applied. This factor, which ranged from 20 to 28 percent, is an estimate of the loss of bollard pull resulting from the motions of the tug in waves. This is merely an estimate, but is in line with similar figures used by others.[7]

The conversion of bollard pull to horsepower must also be an estimate as it varies for each vessel. It depends greatly on the RPM and diameter of the propeller(s) and on whether the tug has Kort nozzle(s) around the propeller(s). For this paper, a factor of 75 horsepower per ton of bollard pull was used. This is typical of a twin screw tug in the 5,000- to 10,000-horsepower class with Kort nozzles.[8] For a 7,000-horsepower tug, this equates to 93 tons of bollard pull.

RESULTS AND CONCLUSIONS

The results of the analysis are displayed in Figures I-9 through I-11. In reviewing them, it is worthwhile to note the major trading routes for various size tankers:

- VLCCs trade primarily to the U.S. Gulf with a smaller number trading ANS Crude from Alaska to the West Coast.
- 140-MDWT tankers trade primarily to the East Coast of the U.S. with a much smaller number trading in the U.S. Gulf and to the West Coast.
- 80-MDWT vessels trade primarily in the U.S. Gulf and on the East Coast with a smaller number trading on the West Coast.

This analysis shows that a tug of 7,000 horsepower is adequate for handling both a VLCC and a 140-MDWT tanker up to just short of a 20-foot (6-meter) sea state and for handling an 80-MDWT tanker up to a sea state of about 21.5 feet (6.5 meters). Beyond that point, a larger tug, or two tugs would be required. By referring to the joint probability curves (Figures I-1 through I-6) one can see that the probability of 6 to 6. 5 meter seas in combination with onshore winds is quite small. For each of the regions, the joint probability of having 16- to 20-foot (5- to 6-meter) waves in conjunction with onshore winds is:

East Coast	0.30%
Gulf Coast	0.05%
California	0.20%
Pacific NW	0.80%
Alaska	1.40%

These numbers represent the percentage of time in certain coastal waters when one would anticipate that a rescue tow or holding operation of a loaded VLCC would be beyond the capability of a single 7,000-BHP tug.

In 1978, DNV published *Towing Operations Guidelines and Recommendations for Barge Transportation*. This document was intended to provide guidance to the offshore industry on how large a tug would be required to transport major equipment offshore. These guidelines recommended using a tug capable of towing in 16.5-foot (5-meter) seas with 39-knot winds and up to a 2-knot current. This correlates very well with the conditions in which we would expect a 7,000-horsepower tug to be able to hold a large tanker.

It is important to note that the wave drift forces increase dramatically up to a 33- to 39-foot (10- to 12-meter) sea state. In these conditions the relative dynamic motions of the tug and tanker would preclude the tug from towing regardless of its horsepower. Even the largest oceangoing tugs cannot continue to tow in major storm conditions.

REFERENCES

1. OCIMF. 1981. *OCIMF* Recommendations on Equipment for the Towing of Disabled Tankers. London: Witherby and Company.
2. Hancox, David. 1987. Reed's Commercial Salvage Practice. Sunderland, England: Thomas Reed Publications Limited.
3. Noel, John V. 1977. Knight's Modern Seamanship, Sixteenth Edition. New York: Van Nostrand Reinhold Company.
4. Hogben, N. et al. 1986. Global Wave Statistics. England: Unwin Brothers Limited.
5. OCIMF. 1992. Users Guide OCMOTA. Maritime Research Institute Netherlands
6. OCIMF. 1977. Prediction of Wind and Current Loads on VLCCs. London: Witherby and Company.
7. Dai, Richard Y. T. et al. 1981. Offshore construction barge performance in towage operations. Proceedings 1981 Offshore Technology Conference. OTC 4164.
8. Blight, Graham J. et al. 1978. Resistance of offshore barges and required tug horsepower. Proceedings 1978 Offshore Technology Conference. OTC 3320.

Wind Speed (knots)	Wave Height–Feet: Average	Wave Height–Feet: Significant Average 1/3 Highest	Wave Height–Feet: Average 1/10 Highest	Average Period	Average Wave Length	Minimum Fetch (nautical miles)	Minimum Duration (hours)	Average Wave Height[b] (maximum)
0	0	0	0	–	–	–	–	–
2	0.05	0.08	0.10	0.5	10 in	5	18 min	
5	0.18	0.29	0.37	1.4	6.7 ft	8	39 min	
8.5	0.6	1.0	1.2	2.4	20	9.8	1.7 hrs	2(3)
10	0.88	1.4	1.8	2.9	27	10	2.4	
13.5	1.8	2.9	3.7	3.9	52	24	4.8	3 1/2(5)
16	2.9	4.6	5.8	4.6	71	40	6.6	
18	3.8	6.1	7.8	5.1	90	55	8.3	
19	4.3	6.9	8.7	5.4	99	65	9.2	6(8 1/2)
20	5.0	8.0	10	5.7	111	75	10	
22	6.4	10	13	6.3	134	100	12	
24.5	8.2	13	17	7.0	164	140	15	9 1/2(13)
26	9.6	15	20	7.4	188	180	17	
28	11	18	23	7.9	212	230	20	
30.5	14	23	29	8.7	258	290	24	13 1/2(19)
32	16	26	33	9.1	285	340	27	
34	19	30	38	9.7	322	420	30	
37	23	37	46.7	10.5	376	530	37	18(25)
40	28	45	58	11.4	444	710	42	
42	31	50	64	12.0	492	830	47	
44	36	58	73	12.5	534	960	52	23(32)
46	40	64	81	13.1	590	1110	57	
48	44	71	90	13.8	650	1250	63	
50	49	78	99	14.3	700	1420	69	29(41)
51.5	52	83	106	14.7	736	1560	73	
54	59	95	121	15.4	810	1800	81	
56	64	103	130	16.3	910	2100	88	37(52)
59.5	73	116	148	17.0	985	2500	101	
>64	>80	>128	>164	18	~	~	~	45(–)

[a] To attain a fully arisen sea for a certain wind speed, the wind must blow at that speed over a minimum distance (fetch) for a minimum time (duration). When winds are 50 knots or more, the required fetch and duration for a fully arisen sea rarely occur. The wave heights shown in the last column, "Average Wave Height" represent what will be found on the average at given wind speeds.

Wave heights refer only to wind waves, and swells from distant or old storms are nearly always superimposed on the wind-wave pattern.

Practical Methods of Observing and Forecasting Ocean Waves, Pierson, Newuman, James, H.O. Pub. 603, 1955.

[b] H.O. 118A

FIGURE I-1 Wind and sea scale for fully arisen sea.[a]

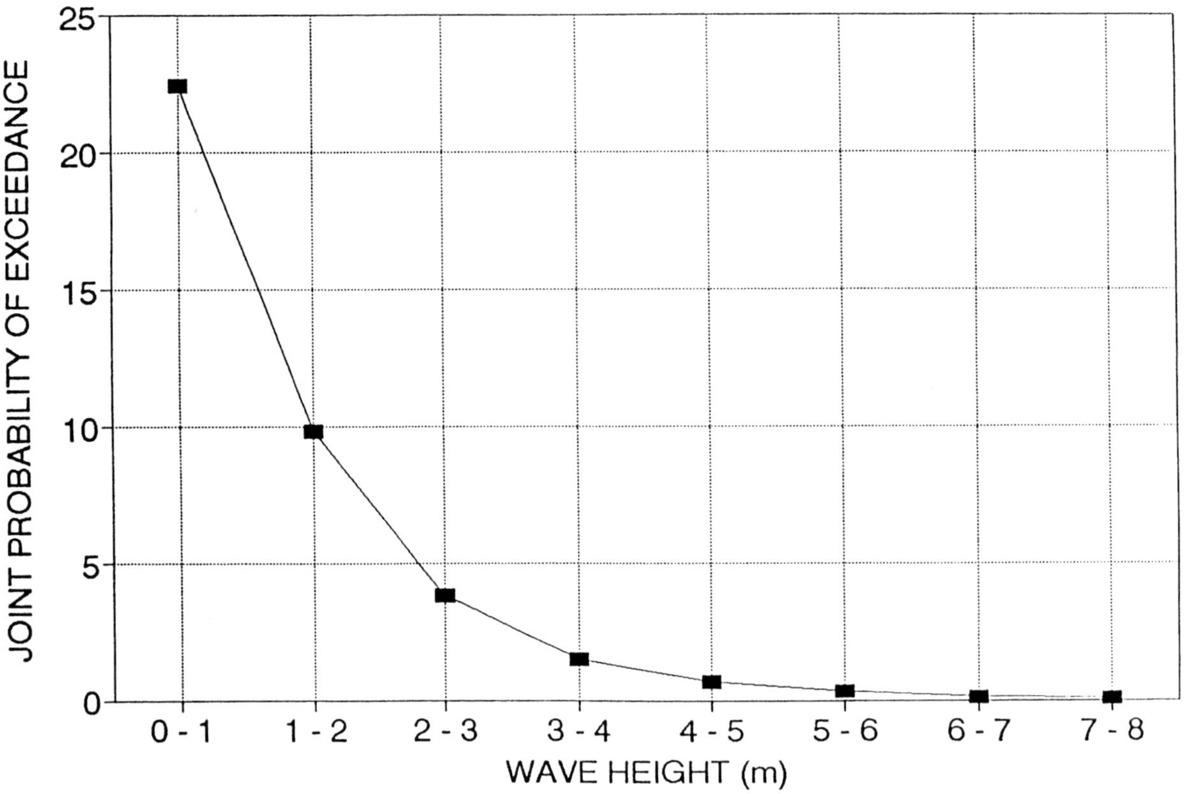

FIGURE I-2 East Coast onshore wave height probability distribution.

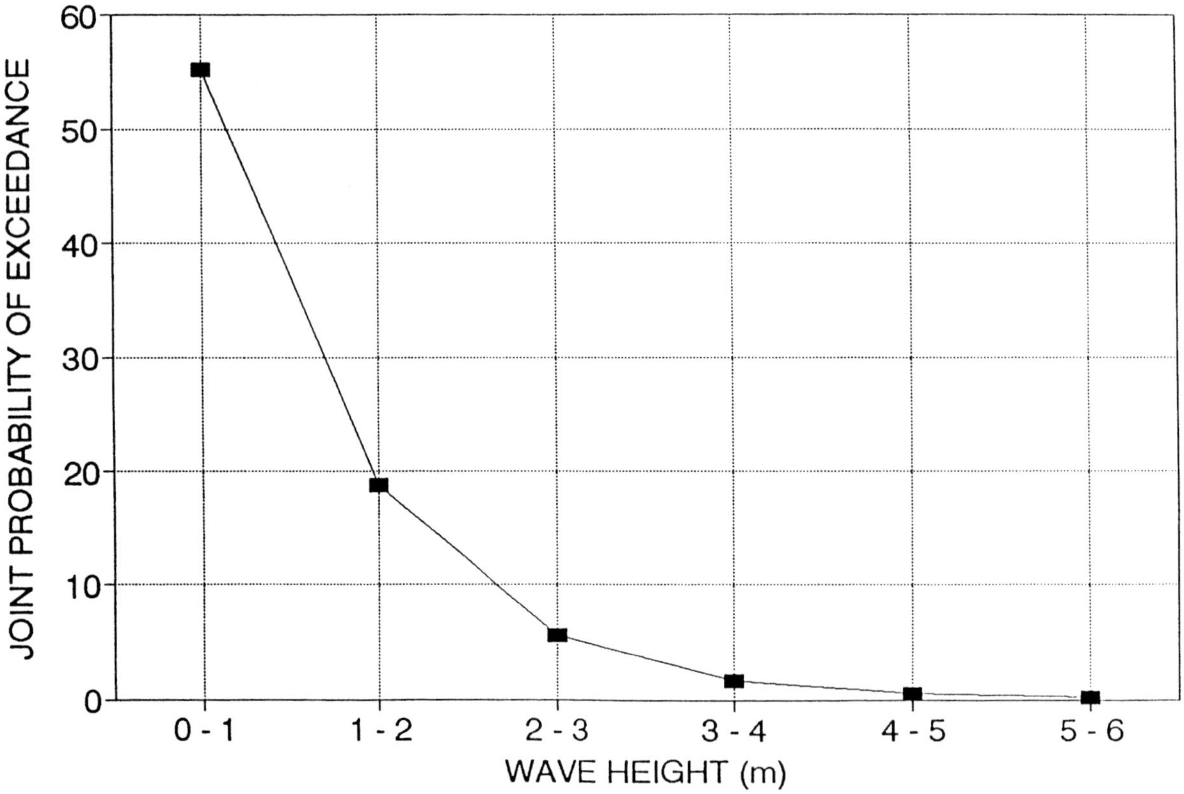

FIGURE I-3 U.S. Gulf Coast onshore wave height probability distribution.

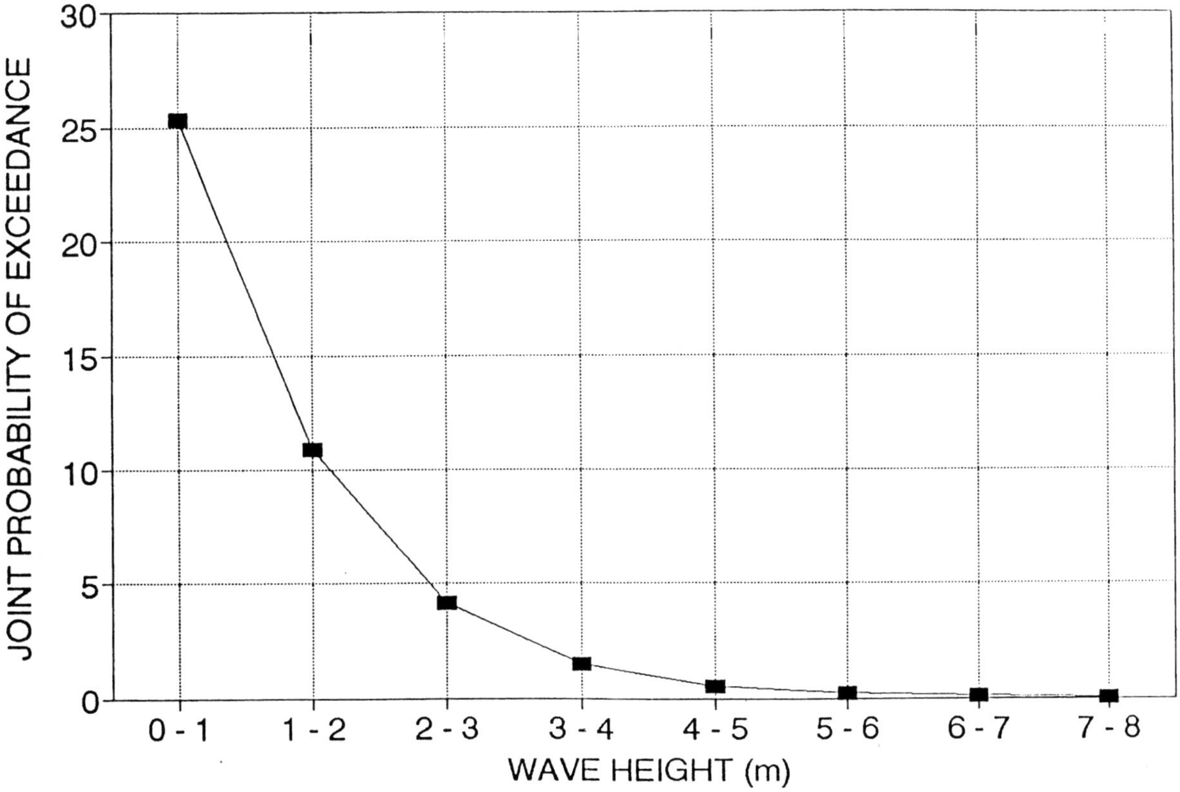

FIGURE I-4 California onshore wave height probability distribution.

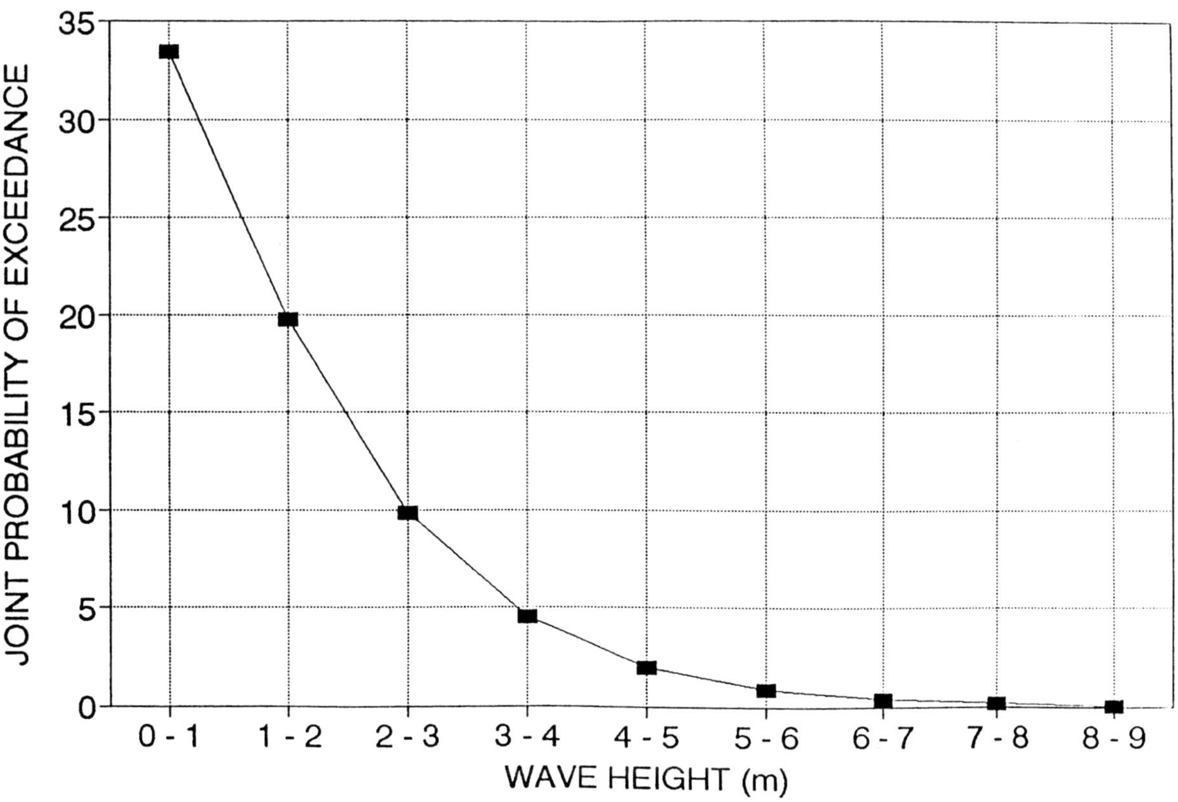

FIGURE I-5 Pacific Northwest onshore wave height probability distribution.

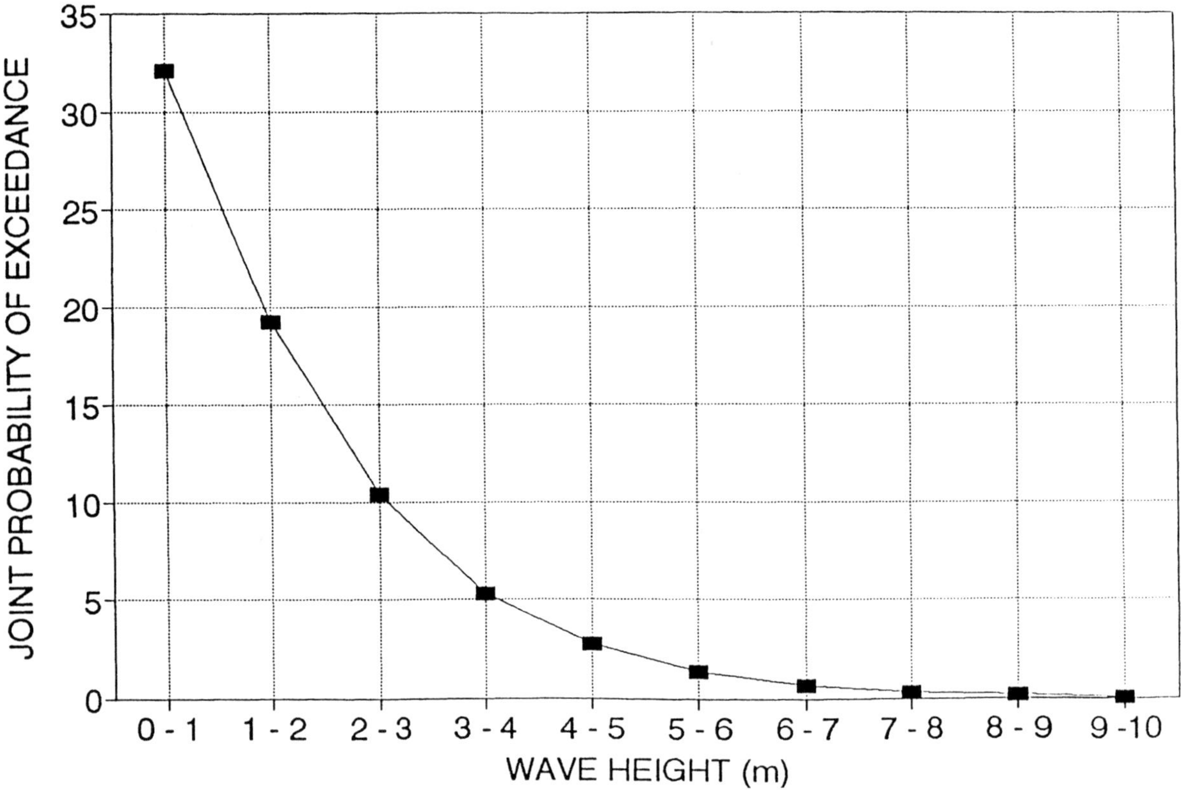

FIGURE I-6 Gulf of Alaska onshore wave height probability distribution.

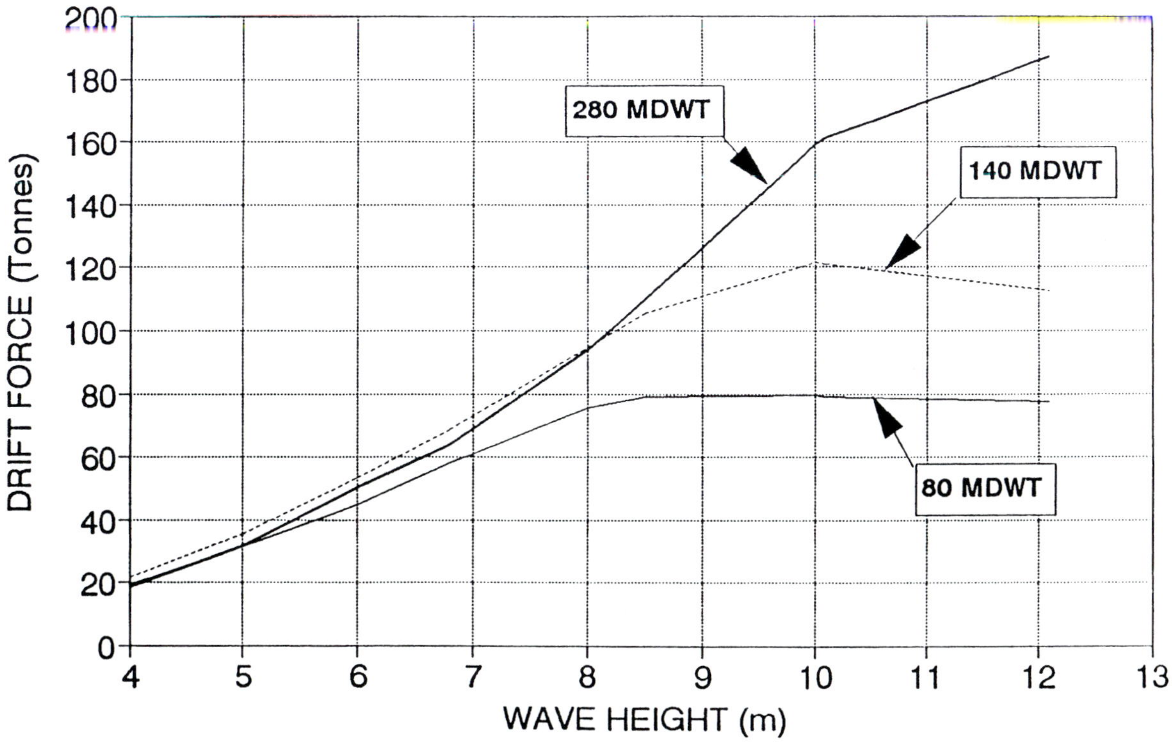

FIGURE I-7 80- to 280-MDWT tanker wave drift force.

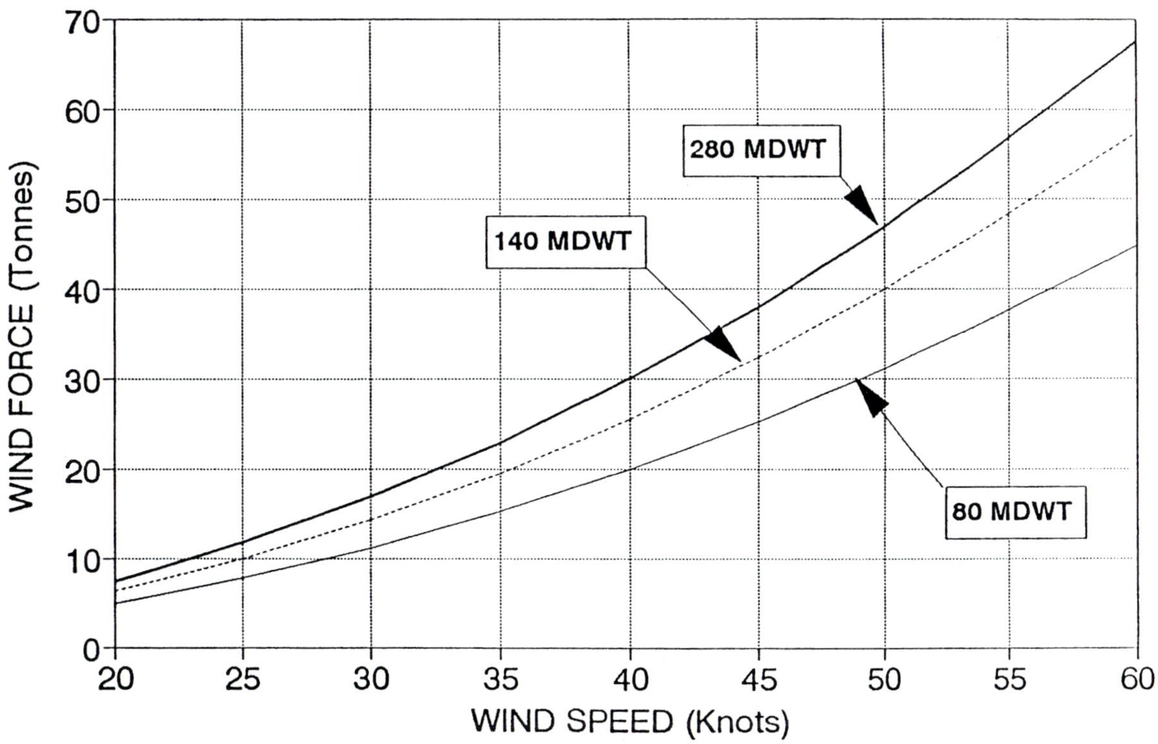

FIGURE I-8 80- to 280-MDWT tanker wind force.

Sig. Wave H	Wind Speed	Tanker Resistance Wind	Tanker Resistance Wave	Tanker Resistance Total	Tug Resist.	Minimum Bollard	Est. Loss of Efficiency	Required Bollard	Estimated Tug BHP
meters	kn	tons	tons	tons	tons	tons	%	tons	HP
4	30	11	20	31	3	34	20	40	3000
5	35	14	31	45	4	49	22	60	4500
6	40	19	45	64	5	69	24	85	6400
7	45	24	76	100	7	107	26	135	10100
10	50	30	79	109	9	118	28	150	11300

Note: Loss of efficiency is estimate to show effect of motions on effective tug power.

FIGURE I-9 Tug power estimate required to hold 80-MDWT fully loaded tanker.

Sig. Wave H	Wind Speed	Tanker Resistance Wind	Tanker Resistance Wave	Tanker Resistance Total	Tug Resist.	Minimum Bollard	Est. Loss of Efficiency	Required Bollard	Estimated Tug BHP
meters	kn	tons	tons	tons	tons	tons	%	tons	HP
4	30	14	22	36	3	39	20	46	3500
5	35	19	36	55	4	59	22	71	5400
6	40	24	54	78	5	83	24	103	7700
7	45	31	95	126	7	133	26	167	12500
10	50	38	121	159	9	168	28	215	16100

Note: Loss of efficiency is estimate to show effect of motions on effective tug power.

FIGURE I-10 Tug power estimate required to hold 140-MDWT fully loaded tanker.

Sig. Wave H	Wind Speed	Tanker Resistance			Tug Resist.	Minimum Bollard	Est. Loss of Efficiency	Required Bollard	Estimated Tug BHP
		Wind	Wave	Total					
meters	kn	tons	tons	tons	tons	tons	%	tons	HP
4	30	16	19	35	3	38	20	46	3400
5	35	22	32	54	4	58	22	71	5300
6	40	29	50	79	5	84	24	104	7800
7	45	36	94	130	7	137	26	173	13000
10	50	45	159	204	9	213	28	272	20400

Note: Loss of efficiency is estimate to show effect of motions on effective tug power.

FIGURE I-11 Tug power estimate required to hold 280-MDWT fully loaded VLCC.

References

Chemical Manufacturers Association (CMA). 1992. Mariner Emergency Response Contingency Plans: An Assessment for U.S. Waters. Washington, D.C.: CMA.

Dean, W. and L.L. Crick. 1983. New laws affecting the jettisoning of oil. In Purposeful Jettison of Petroleum Cargo. Washington, D.C.: National Academy Press.

Disabled Tanker Towing Study Group. 1993. Prince William Sound Disabled Tanker Towing Study Part I: Evaluation of Existing Equipment, Personnel and Procedures. Anchorage, Alaska.

Edgar, K. 1993. Presentation to the Committee on Marine Salvage Issues, New York City, August 3, 1993.

Ellis, M. 1994. Significance of jettisoning. In Purposeful Jettison of Petroleum Cargo. Washington, D.C.: National Academy Press.

Engelhardt, R. 1994. Importance of spill size on the environmental effects of a spill incident. In Purposeful Jettison of Petroleum Cargo. Washington, D.C.: National Academy Press.

Galt, J. 1994. Oil spill trajectory modeling. In Purposeful Jettison of Petroleum Cargo. Washington, D.C.: National Academy Press.

International Chamber of Shipping and Oil Companies International Marine Forum (OCIMF). 1992. Peril at Sea and Salvage: A Guide for Masters. 4th Ed. London: Witherby and Co. Ltd.

Johnson, J. 1993. Presentation to the Committee on Marine Salvage Issues, New York City, August 3, 1993.

Johnson, R. 1993. The National Response System: Where do we go from here? Pp. 67-72 in Proceedings of the 1993 International Oil Spill Conference (Prevention, Preparedness, Response). Washington, D.C.: American Petroleum Institute.

Milwee Associates Inc. 1993. Precedents for subsidized salvage stations. Paper prepared for Committee on Marine Salvage Issues. Portland, Oregon: Milwee.

National Research Council, 1982. Marine Salvage in the United States. Washington, D.C.: National Academy Press.

National Research Council. 1985. Oil in the Sea. Washington, D.C.: National Academy Press.

National Research Council. 1986. An Assessment of External Firefighting for Marine Structures. Washington, D.C.: National Academy Press.

National Research Council. 1991. Tanker Spills: Prevention by Design. Washington, D.C.: National Academy Press.

National Research Council. 1994. Purposeful Jettison of Petroleum Cargo. Washington, D.C.: National Academy Press.

Office of Technology Assessment. 1990. Coping with an Oiled Sea. Washington, D.C.: U.S. Government Printing Office.

Nicholas, R.H., Jr. 1993. New laws governing the jettisoning of oil. In Purposeful Jettison of Petroleum Cargo. Washington, D.C.: National Academy Press.

Oil Companies International Marine Forum (OCIMF). 1981. Recommendations on Equipment for the Towing of Disabled Tankers. Oil Companies International Marine Forum. London, England.

Porricelli, J.D., J.H. Boyd, Jr., and K.E. Schleiffer. 1983. Modern Analytic Techniques in Salvage Engineering Using Portable Computers. New York: Society of Naval Architects and Marine Engineers.

Rees, V. Lee Okarma. 1994. Salvaging a coherent approach to pollution response: A comment on new federal and state laws that govern the jettisoning of cargo during salvage operations. In Purposeful Jettison of Petroleum Cargo. Washington, D.C.: National Academy Press.

Ringelberg, J. and D. Banks. 1993. Presentation to the Committee on Marine Salvage Issues, New York City, August 2-3, 1993.

Smith, P. and G.H. Reed. 1990. Salvage response in the U.S. Paper presented at the 11th International Tug Convention and International Marine Salvage Symposium, Halifax, Nova Scotia. September 27, 1990.

Stiehl, G. 1993. Presentation to the Committee on Marine Salvage Issues, Seattle, Washington, June 3, 1993.

Sweeney, J. 1993. Presentation to the Committee on Marine Salvage Issues, New York City, August 2-3, 1993.

Tecnitas. 1992. International Salvage Industry Survey. Paris: Bureau Veritas. April.

U.S. Coast Guard. 1993. Deepwater Port Study. Washington, D.C.

White, R. 1993. Presentation to the Committee on Marine Salvage Issues, New Orleans, August 9, 1993.

Williams, L. 1993. Presentation to the Committee on Marine Salvage Issues, New Orleans, August 9, 1993.